# PROGRAMMING AND INTERFACING ATMEL® AVR® MICROCONTROLLERS

## THOMAS GRACE

**Cengage Learning PTR**

CENGAGE
Learning®

Professional • Technical • Reference

Australia • Brazil • Mexico • Singapore • United Kingdom • United States

**CENGAGE Learning**

Professional • Technical • Reference

**Programming and Interfacing ATMEL® AVR® Microcontrollers**

**Thomas Grace**

**Publisher and General Manager, Cengage Learning PTR:** Stacy L. Hiquet

**Manager of Editorial Services:** Heather Talbot

**Product Team Manager:** Emi Smith

**Project Editor/Copyeditor:** Dan Foster

**Technical Editor:** Joseph Ozvold

**Interior Layout Tech:** MPS Limited

**Cover Designer:** Mike Tanamachi

**Indexer:** Valerie Haynes Perry

**Proofreader:** Jenny Davidson

**Cover images:** © Lukas Rs/ Shutterstock.com

For product information and technology assistance, contact us at **Cengage Learning Customer & Sales Support, 1-800-354-9706**.

For permission to use material from this text or product, submit all requests online at **cengage.com/permissions**.

Further permissions questions can be emailed to **permissionrequest@cengage.com**.

Atmel and AVR are registered trademarks of Atmel Corporation or its subsidiaries, in the US and/or other countries.

All images © Cengage Learning PTR unless otherwise noted.

Library of Congress Control Number: 2015934729

ISBN-13: 978-1-305-50999-3

ISBN-10: 1-305-50999-4

**Cengage Learning PTR**

20 Channel Center Street

Boston, MA 02210

USA

Cengage Learning is a leading provider of customized learning solutions with employees residing in nearly 40 different countries and sales in more than 125 countries around the world. Find your local representative at **www.cengage.com**.

Cengage Learning products are represented in Canada by Nelson Education, Ltd.

For your lifelong learning solutions, visit **cengageptr.com**.

Visit our corporate website at **cengage.com**.

Printed in the United States of America

Print Number: 01                    Print Year: 2015

*For Evan, Cassie, and Patti*

## Acknowledgments

I would like to thank the editorial staff, and especially Joseph Ozvold, for all their time and help in bringing this book together.

## About the Author

**Thomas Grace** earned B.S and M.S. degrees in Electrical and Computer Engineering from Clarkson University. Over the years, Grace has worked for IBM in Burlington, VT; Indiana University-Purdue University Indianapolis at PLSP in Batu Pahat, Malaysia; University at Buffalo at the Institute of Technology MARA in Kuala Lumpur, Malaysia; and Clarkson University. For over 25 years he has taught engineering at Broome Community College in Binghamton, NY, and owns a small electronics manufacturing company. Outside of work he enjoys time with his family, traveling, and woodworking.

# CONTENTS

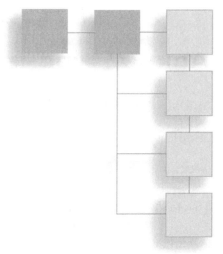

# INTRODUCTION

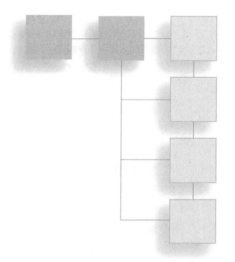

*Programming and Interfacing ATMEL AVR Microcontrollers* introduces you to ATMEL's AVR microprocessors—the same chip used in the Arduino. You'll learn how to use the AVR to interact with devices such as motors, LCD screens, GPS sensors, touch pads, temperature sensors, accelerometers, distance sensors, and other interesting hardware. Other topics covered include analog to digital conversion, digital I/O, interrupts, serial peripheral interface, and serial communications.

In addition, this book shows you how to use AVR Studio to program the ATtiny13, ATmega328, and ATmega32 in assembly, C, and C++ languages.

The book is divided into four chapters:

- **Chapter 1, "Digital Systems,"** introduces the background information needed to program a microprocessor. LEDs, switches, and logic gates are also covered in this chapter.

- **Chapter 2, "AVR Programming,"** shows you how to program the AVR first in assembly, then in C, and then in C++.

- **Chapter 3, "Hardware Interfacing,"** concentrates on how to interface various hardware devices such as sensors, motors, and displays to an AVR.

- **Chapter 4, "Projects Using the AVR,"** provides example projects. You can build these projects as shown or perhaps use them as a starting point for your own projects. Some of the projects here are complete, while others are presented in various stages of completion.

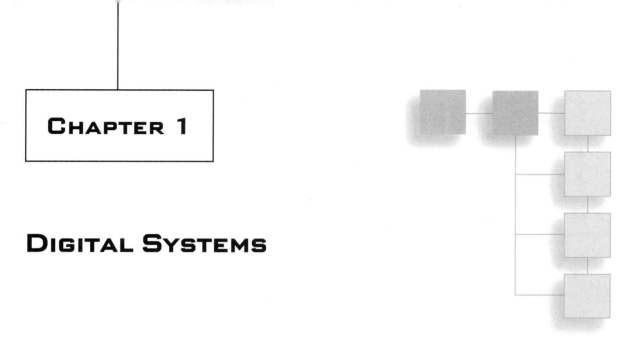

# CHAPTER 1

# DIGITAL SYSTEMS

The microprocessor is a digital system. It is made up of logic gates, and thus its native number system is binary. Microprocessors are very often connected to LEDs, switches, and logic gates, which are covered later in this chapter. Signals in a digital system are either a high state or a low state. This chapter will also discuss the defined voltages for each state.

## NUMBER SYSTEMS

In order to program the microprocessor, we must fully understand binary and hex numbers. In addition, we need to know how to convert them to and from base 10, since the architecture of the microprocessor is based on binary and because we use base 10. Hex is used because it is a convenient way of writing a binary number.

### Base 10 (Decimal)

Roman numerals were used in Europe until Fibonacci introduced Western Arabic numerals by way of Africa into Italy, thus making financial calculations much easier. Nearly everyone uses base 10, perhaps because we have 10 fingers. The 10 digits are 0, 1, 2, 3… 9. There is no single digit that represents 10. In other words, 9 is the highest single digit in base 10, and when you add 1 you must carry to the next column. A value of ten in the first column is carried over to get a 1 in the second column. These rules, which you learned years ago, apply to other bases.

It is important to recall how to work with decimal numbers. What is $842_{10}$ equal to? Answer: $8 \times 10^2 + 4 \times 10^1 + 2 \times 10^0$.

## Roman Numerals

There are seven roman numerals (I = 1, V = 5, X = 10, L = 50, C = 100, D = 500, M = 1000) but interestingly no zero. An XII represents 10 + 2 by an additive process. When a lower-value letter is placed in front of a higher-value letter, a subtractive process is used. For example, IX equals 10 − 1, or 9. Imagine the difficulty when given Roman numbers to add, subtract, multiply, or divide. For example, what is II percent of LIX?

## Base 2 (Binary)

In binary, there are two digits: 0 and 1. Let's count: 0, 1, 10, 11, 100, 101, 110, 111. We just counted from 0 to 7. Computers use binary as their number system because transistors have two states: on and off. One bit is a single 1 or 0 in binary. 4 bits constitute a nibble. 8 bits make a byte. Often, we section off (with a space) large binary numbers into bytes to make them readable and less prone to errors in transcription.

## Base 8 (Octal)

In octal, there are eight digits (0, 1, 2… 7). Counting in octal follows: 0, 1, 2, 3, 4, 5, 6, 7, 10, 11, 12. The number $13_8$ is equal to $1 \times 8 + 3 \times 1$ or $11_{10}$. This base is infrequently used.

## Base 16 (Hexadecimal)

Base 16, or hexadecimal, is formed from the two terms hex(6) and decimal(10). Hexadecimal is typically shortened to the term hex.

In hex, there are 16 digits (0, 1, 2, 3…8, 9, A, B, C, D, E, F). We use letters to help create the extra six new digits. Thus A has the value of 10 (decimal), B is 11, etc. The number 12 in hex is equal to $16 + 2 = 18_{10}$. What is $842_{16}$ equal to? $8 \times 16^2 + 4 \times 16^1 + 2 \times 16^0 = 2114_{10}$.

## Base 60 (Sexagesimal)

This base originated with the ancient Sumerians in the third millennium BC, and passed down to the ancient Babylonians. With a base of 60 (and 60 characters to remember), perhaps this is why the Babylonian civilization faded away. More importantly, it was used in a modified form to measure time and angles.

## Counting

Remember that when counting there is a carry over to the next column when the maximum value is reached for that base. It is good practice to state the base of a number, because the base is unknown if it is not stated! Some of the numbers listed in the following table could be a valid number in another base.

| Base 10 | Base 2 | Base 16 | Roman |
|---------|--------|---------|-------|
| 0 | 0000 | 0 | |
| 1 | 0001 | 1 | I |
| 2 | 0010 | 2 | II |
| 3 | 0011 | 3 | III |
| 4 | 0100 | 4 | IV |
| 5 | 0101 | 5 | V |
| 6 | 0110 | 6 | VI |
| 7 | 0111 | 7 | VII |
| 8 | 1000 | 8 | VIII |
| 9 | 1001 | 9 | IX |
| 10 | 1010 | A | X |
| 11 | 1011 | B | XI |
| 12 | 1100 | C | XII |
| 13 | 1101 | D | XIII |
| 14 | 1110 | E | XIV |
| 15 | 1111 | F | XV |
| 16 | 10000 | 10 | XVI |

## Converting Base 2 to Base 10

For a base 2 number, the value of the first digit left of the decimal point is 1, the next column has a value of 2, and the other columns have a value of: 4, 8, 16, 32,...etc. To the right of the decimal point, the values of each column are half that of the column to the left. Thus, the weights of each digit appear as shown here: ...32, 16, 8, 4, 2, 1, 1/2, 1/4, 1/8, 1/16...etc.

Examples:

| Base 2 | Base 10 |
|---|---|
| 1101 | $= 8 + 4 + 0 + 1 = 13$ |
| 1010111 | $= 64 + 0 + 16 + 0 + 4 + 2 + 1 = 87$ |
| 1010.0111 | $= 8 + 2 + 1/4 + 1/8 + 1/16 = 10.4375$ |

## Converting Base 10 to Base 2

The process of converting decimal (base 10) to binary (base 2) is to take the base 10 number and continually subtract numbers that are powers of 2 from highest to lowest (i.e., 64, 32, 16, 8, 4, 2, 1).

If, for example, you are able to subtract 16 from the base 10 number, then the fourth binary bit is set as a 1; otherwise, it is 0.

Next, if you can subtract 8, the third binary bit is 1; otherwise, it is 0.

Next, if you can subtract 4, the second binary bit is 1; otherwise, it is 0.

Next, if you can subtract 2, the first binary bit is 1; otherwise, it is 0.

Next, if you can subtract 1, the zeroth binary bit is 1; otherwise, it is 0.

Example: Convert 27 decimal to base 2:

```
27 =>    _   _   _   _   _   _
        32 16   8   4   2   1
```

Start from the left and place a 1 if the binary weight fits within the base 10 number.

```
27–32= no, subtract  0
27–16= 11, subtract  1
11– 8=  3, subtract  1
 3– 4= no, subtract  0
 3– 2=  1, subtract  1
 1– 1=  0, subtract  1
```

Thus, the base 2 result is $011011_2$

Other examples:

```
11₁₀   => 11–8=3,3–2=1,1–1=0    => 1011₂
12₁₀   => 12–8–4=0              => 1100₂
312₁₀ => 312–256–32–16–8 = 0   => 100111000₂
```

## Converting Base 2 to Base 16

For a base 2 number, the value of the first digit left of the decimal point is 1, the next column is 2, and the next columns have values of: 4, 8, 16, 32,...etc. To the right of the decimal point, the values of each column are: 1/2, 1/4, 1/8,...etc. The way to make the conversion into hex is to group the binary number into nibbles (sections of 4 bits). Each binary nibble corresponds to one digit in hex. The reason for this is that $2^4 = 16^1$, a 4:1 ratio. One can pad with zeros on the far left or far right of the decimal point without changing the value of the binary number. Following are examples of converting base 2 to base 16.

| Base 2 | Base 16 |
| --- | --- |
| 1001 | 9 |
| 10 1010 | 2A |
| 1101 | D |
| 101 0111 | 57 |
| 1111 1010.0111 1 | FA.78 |

## Converting Base 16 to Base 2

To convert base 16 to base 2, just reverse the process by converting 1 hex digit into 4 binary digits. Following are examples of converting base 16 to base 2.

| Base 16 | Base 2 |
| --- | --- |
| IFA | 0001 1111 1010 |
| 23.1 | 0010 0011.0001 |
| ABC1.9 | 1010 1011 1100 0001.1001 |

## Converting Base 10 to Base 16

One way to do this conversion is to do it via base 2. In other words, convert base 10 to base 2, and then to base 16.

For example, converting $137_{10}$ to hex is:

$= 128 + 8 + 1$ decimal

$= 1000\ 1001$ binary

$= 89$ hex

## Binary Coded Decimal

The Binary Coded Decimal (BCD) number system uses four binary digits to represent one decimal digit. This means that binary values above $1001_2$, or 9, are not used. The reason for using BCD in a computer is that the conversion between BCD and decimal is easier than the conversion between binary and decimal. The downside is that BCD uses more memory space than binary because the combinations 1010 to 1111 are not used. BCD numbers are used with seven-segment displays, as seen in such things as clocks, vending machines, and calculators.

To convert BCD to decimal: Group each four digits together and convert the group to a decimal digit.

| BCD | Decimal |
| --- | --- |
| 1001 | 9 |
| 0111 | 7 |
| 1100 | anything over 9 is invalid BCD |
| 11 0001 | 31 |
| 101 0110 | 56 |

To convert decimal to BCD: Take each digit and make a group of four digits. Use a space after every fourth digit to help see the answers.

| Decimal | BCD |
| --- | --- |
| 921 | 1001 0010 0001 |
| 32 | 0011 0010 |
| 101 | 0001 0000 0001 |

Again, each group of four binary digits makes a number from 0 to 9. This makes it convenient to work with digital displays. Use BCD when working with seven-segment displays and storage is not a problem. The invalid BCD numbers make this somewhat inefficient in terms of storage. An 8-bit BCD number would range from 0 to 99, while an 8-bit base 2 number can range from 0 to 255. This means that 8-bit BCD numbers utilize 100 of 256 possibilities, or 39%. The advantage for BCD is that computer input and output of numbers is sometimes easier than binary.

## Base 2: Adding and Subtracting

The rules for adding and subtracting binary numbers are the same rules as in base 10. We have used base 10 for so long that it is second nature to us, but think of how you add and subtract in base 10 and use the same procedure for binary.

In binary, the highest digit is a 1. If you add up a column and get more than 1 (a 2 or 3), then that 2 or 3 has to go to the next column to the left as a carry, which is a 1. Here, there is a carry from bit 1 to bit 2. The rightmost column is called bit zero.

```
  00010
+ 0011
= 0101
```

Lots of carries here:

```
  00101010101
+   0010111011
= 01000010000
```

Try subtraction in base 10. Do it slowly and show the borrow. The 9 becomes an 8. You carry 10 over because it is base 10. 10 is carried to the first column, making 11. Now take 6 from 11 to get 5.

```
  91
-  6
= 85
```

Back to binary. When you borrow a 1 from the left it becomes a value of 2 in the column to the right because it is base 2. Here is the example of $2-1$, which is 1. The 1 in bit position 1 is borrowed over into bit position 0 and is written as 2. Now subtract.

```
  010
- 01
= 01
```

For subtraction, start on the right again and watch the borrows.

```
  0111
-  011
= 0100
```

Another example:

```
  010101
-  00111
= 01110
```

To check your work, convert all these numbers to base 10.

## Base 2: Signed and Unsigned Numbers

Computers can store only 1s and 0s and not a negative sign. To represent a negative number, the twos complement will be used. Keep in mind that the programmer has to keep track if the data being used is signed ($+/-$ numbers) or unsigned (just $+$ numbers). Suppose you see the number $11111111_2$ in memory. What does it represent? If it is unsigned, then the value is 255, but if it is a signed number, the value is $-1$. So one cannot look at a number and determine whether it is a signed or unsigned number.

### Unsigned Numbers

Here is the 8-bit unsigned number line. There are 256 numbers.

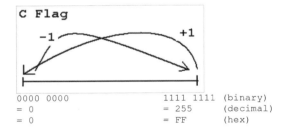

```
C Flag

   -1                      +1



0000 0000           1111 1111  (binary)
= 0                 = 255      (decimal)
= 0                 = FF       (hex)
```

Unsigned numbers represent only zero and the positive numbers. The largest number would be all ones. Consider 8-bit unsigned numbers. They would have a range of 0 to 255. Using 16 bits unsigned, the range is from 0 to 65535. What would happen if you took the largest number and added 1? Answer: You would get zero (the smallest number). Or, if you took the smallest number and subtracted 1, you would get the largest number (all ones). When you go from all zeros to all ones or vice versa, a carry is generated. From this discussion, you can see that computers cannot count to infinity, and if the programmer is not careful, you might report that 255 + 1 is zero!

### Signed Numbers

Here is the 8-bit signed number line. There are 256 numbers.

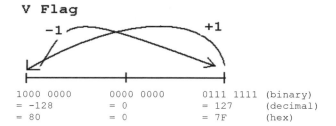

```
V Flag
    -1                          +1



1000 0000      0000 0000      0111 1111  (binary)
= -128         = 0            = 127      (decimal)
= 80           = 0            = 7F       (hex)
```

Signed numbers have a range such that half the numbers are negative and half are positive. If you are working with signed numbers, then the most significant bit (MSB) indicates the sign. If the MSB is 0, it is positive; if the MSB is 1, it is negative. For example, let's work with 8-bit signed numbers and count from zero: $00000000_2$, 00000001, 00000010, ...$01111111_2$. At this point, we have counted from 0 to 127. If we go one higher, we get $100000000_2$. But this must be a negative number because the MSB is 1. Now start from zero and count down: $00000000_2$, 11111111, 11111110, ...$10000000_2$. We just counted from 0, −1, −2 to ... −128. Notice that $11111111_2$ has the value of −1. These numbers wrap around. If you take −128 ($10000000_2$) and subtract 1, you get 127 ($01111111_2$) which, of course, is incorrect. When going from $10000000_2$ to $01111111_2$ or vice-versa, an overflow is generated, not a carry.

### Two's Complement

The two's complement will take the negative of a signed number. To do this, take the binary number, toggle all the bits, and then add one. The two's complement of 1001 is 0110 + 1 = 0111 = 7. So, 1001 is −7 because 0111 is +7.

Following are some examples using 8-bit signed numbers.

1. 19 as an 8-bit signed number is simply: $19 = 00010011_2$.

2. To find −19 as an 8-bit signed number, start with $19 = 00010011_2$, toggle all bits => 1110 1100, and add 1 => $1110 1101_2$, which is −19.

3. To write −1 as an 8-bit signed number, start with $1 = 0000 0001_2$, toggle all bits => 1111 1110, add 1 => $1111 1111_2$, and this is −1.

4. As a signed number, what does 1100 0011 represent? Since the MSB is a 1, it is negative. To find the value, take the two's complement. Original number: 1100 0011 Toggled: 0011 1100, and adding one yields 0011 1101, which is +61. So, we started with −61.

## Range of Numbers

As you can see, using 8 bits unsigned yields a range of numbers from all zeros to all ones, which in decimal is 0 to 255.

| Number of Bits | Signed/Unsigned | Range of Numbers |
|---|---|---|
| 6 | Unsigned | 0 to $2^6-1 = 63$ |
| 8 | Unsigned | 0 to 255 |
| 10 | Unsigned | 0 to $2^{10}-1 = 1023$ |
| $n$ | Signed | $-2^{n-1}$ to $2^{n-1}-1$ |
| 8 | Signed | −128 to 127 |
| 3 | Signed | −4 to 3 |
| 10 | Signed | −512 to 511 |

## Base 16: Addition and Subtraction

The rules for working in hex are the same rules as in base 10. We have used base 10 for so long that it is second nature to us, but think about how you add and subtract in base 10 and do the same for hex. In hex, the highest digit is an F. If you add up a column and get more than F (15), then 16 must go to the next column to the left as a carry (a carry is a 1).

In this example, there is a carry from column 2 to column 3.

```
000F1
+001A
=010b (hex)
```

Lots of carries in this next example:

```
0AC43
+2C3F
=0D882 (hex)
```

Try subtraction in base 10. Do it slowly and show the borrow. In the example, the 9 becomes an 8. You borrow 10 over because it is base 10. 10 is borrowed to the first column making an 11. Now take 6 from 11 to get 5.

```
 91
 -6
=85 (decimal)
```

Back to hex. When you borrow a 1 from the left it becomes a value of 16 in the column to the right because it is base 16. Here is the example of 10−1, which is F. The 1 in column 2 is borrowed over into column 1 and is written as 16. Now subtract.

```
010
-01
00F (hex)
```

For subtraction, start on the right again and watch the borrows.

```
 07F1
 -212
=05DF
```

To check your work, you can convert all these numbers to base 10, or in the case of subtraction add the last two numbers.

## DIGITAL LOGIC

Logic gates are the building blocks of computers and other digital devices. Millions of gates can be placed onto an integrated circuit (IC) or chip. The logic gate has one or more inputs and one or more outputs. Digital signals are either high or low. Typically, a high is 5v (sometimes 3.3v) and a low is 0v. In actuality, the voltage levels may vary from these ideal conditions of 5 and 0 volts. If, for example, the load connected to the output of a gate draws too much current, then the voltage level can vary from the ideal and thus errors can occur. Digital signals have two states. These two states are referred to by several different names, which are:

5v = high = true = 1 = on

0v = low = false = 0 = off

Any voltage between 5v and 2.0v going into a gate is considered a high, while any voltage between 0.8v and 0v is considered a low. Logic gates make decisions based on the inputs and then in turn can drive an output device. For example, an alarm is to go off when any one of the four doors of a car is open while the car engine is on.

alarm = (door1 OR door2 OR door3 OR door4) AND engine

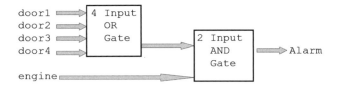

## Logic Gates

In the following, we will use the family of logic gates called TTL, which is short for Transistor-Transistor Logic. Specifically, we will use the 7400 series.

Basic gates include the logic functions: AND, OR, NAND, NOR, EXOR, NOT. The following integrated circuits generally contain four or six logic gates. The letters a and b represent inputs to the gate, while the letter y is the name of the gate output.

| Part # | Description |
| --- | --- |
| 7400 | 2 input NAND, inverse of AND |
| 7402 | 2 input NOR, inverse of OR |
| 7404 | 1 input NOT, inverts signal |
| 7408 | 2 input AND, both inputs must be true for the output to be true (y = ab) |
| 7432 | 2 input OR, if either input is true, the output will be true |

| a | b | y |
| --- | --- | --- |
| 0 | 0 | 0 |
| 0 | 1 | 0 |
| 1 | 0 | 0 |
| 1 | 1 | 1 |

OR, 7432 - Output is true when any input is true. Equation: y = a+b.

| a | b | y |
|---|---|---|
| 0 | 0 | 0 |
| 0 | 1 | 1 |
| 1 | 0 | 1 |
| 1 | 1 | 1 |

NOT, 7404 - Inverts signal. Equation: y = /a.

| a | y |
|---|---|
| 0 | 1 |
| 1 | 0 |

NAND, 7400 - Inverse of AND. Equation: y = /(ab).

| a | b | y |
|---|---|---|
| 0 | 0 | 1 |
| 0 | 1 | 1 |
| 1 | 0 | 1 |
| 1 | 1 | 0 |

NOR, 7402 - Inverse of OR. Equation: y = /(a+b).

| a | b | y |
|---|---|---|
| 0 | 0 | 1 |
| 0 | 1 | 0 |
| 1 | 0 | 0 |
| 1 | 1 | 0 |

EXOR, 7486 - Exclusive OR. True when only one input is true. Equation: $y$ = a exclusively ORed with b.

| a | b | y |
|---|---|---|
| 0 | 0 | 0 |
| 0 | 1 | 1 |
| 1 | 0 | 1 |
| 1 | 1 | 0 |

There are many other gates, but all can be built by some combination of the above simple gates. For example, to get a four-input OR gate, you could use three two-input OR gates. Each of the above gates has one output, which will be either a high (5v) or a low (0v). To see the output, you can connect it to a voltmeter or an LED. Connect switches to the inputs of the gates. Do not forget that these devices need power to operate. Connect 5v to the Vcc pin (usually pin 14) and 0v to the GND pin (usually pin 7). The pins are numbered starting at the notch and going counter clockwise.

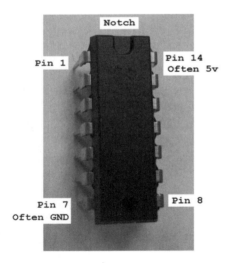

A single chip often contains several gates. These pin-outs show which pins are input, output, and power.

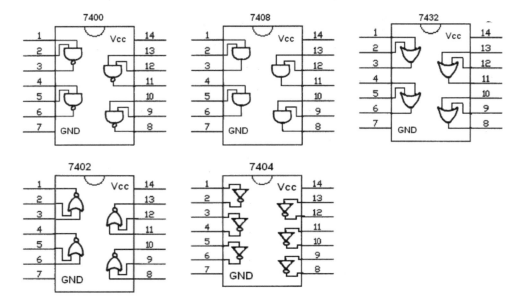

## Combinational Logic

In this section, we consider combining the gates to do some sort of larger function. Every gate has some small delay of approximately a few nanoseconds—typically less than 10 ns. Thus, the output of the circuit will occur very soon after the input but will take some time, which can cause problems. For a combinational logic circuit, if an input changes, the output will be updated automatically and without a clock. A sequential logic circuit, on the other hand, does require a clock to move the data through the circuit.

Consider this example of using gates: A fire alarm is to go off when any one of the four sensors is active and the alarm is activated. Each of the sensors goes to one of the input lines of the OR gates. The activation switch would go to the bottom input on the AND gate. The output of the AND gate might go to a relay to sound the alarm (not shown). If each gate takes 10 ns, then the time delay for the entire circuit is approximately 30 ns.

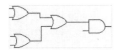

This example shows how to determine and simplify the equation of a circuit.

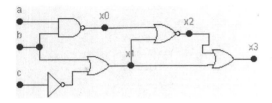

The equation follows. The "/" represents a NOT gate.

x0 = /(ab)

xl = b+/c

x2 = /(x0+xl)

x3 = x2+xl

x3 = /(/(ab)+(b+/c)) + b+/c

x3 = (ab/(b+/c)) + b+/c

x3 = (ab/bc) + b+/c

x3 = 0+b+/c

x3 = b+/c

The reduced equations show that the larger circuit can be reduced to one OR and one NOT gate. The following truth table lists all possible input combinations and the corresponding output.

| a | b | c | x3 |
|---|---|---|----|
| 0 | 0 | 0 | 1 |
| 0 | 0 | 1 | 0 |
| 0 | 1 | 0 | 1 |
| 0 | 1 | 1 | 1 |
| 1 | 0 | 0 | 1 |
| 1 | 0 | 1 | 0 |
| 1 | 1 | 0 | 1 |
| 1 | 1 | 1 | 1 |

What is the circuit and truth table for y = /(CA)B?

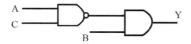

| a | b | c | y |
|---|---|---|---|
| 0 | 0 | 0 | 0 |
| 0 | 0 | 1 | 0 |
| 0 | 1 | 0 | 1 |
| 0 | 1 | 1 | 1 |
| 1 | 0 | 0 | 0 |
| 1 | 0 | 1 | 0 |
| 1 | 1 | 0 | 1 |
| 1 | 1 | 1 | 0 |

### Sum of the Products (SOP)

One method of going from the truth table to an equation is to use sum-of-the-products (SOP). For every 1 in the output (y) column, there will be one term in the equation that is based on the input variables. Each of these terms will produce a 1 and they will be ORed together. For the truth table below, the SOP equation will have three terms. The first term (from row 1) is one when a is zero, b is zero, and c is zero. The SOP does not guarantee simplest form and is usually far from it. The second term comes from row 5, and the third term comes from row 6.

| a | b | c | y |
|---|---|---|---|
| 0 | 0 | 0 | 1 |
| 0 | 0 | 1 | 0 |
| 0 | 1 | 0 | 0 |
| 0 | 1 | 1 | 0 |
| 1 | 0 | 0 | 1 |
| 1 | 0 | 1 | 1 |
| 1 | 1 | 0 | 0 |
| 1 | 1 | 1 | 0 |

SOP Equation: y = /a/b/c + a/b/c + a/bc

## Product of the Sums (POS)

Another method of going from the truth table to an equation is to use product-of-the-sums (POS). For every 0 in the output column, there will be one term in the equation that covers it based on the input variables. Each of these terms will produce a zero, and they will be ANDed together. So this works well when there are fewer zeros than ones. For the truth table below, the POS equation will have three terms. The first term (from row 4) is zero when a is zero, b is one, and c is one. The POS does not guarantee simplest form and is usually far from it. The second term comes from row 7, and the third term comes from row 8.

| a | b | c | y |
|---|---|---|---|
| 0 | 0 | 0 | 1 |
| 0 | 0 | 1 | 1 |
| 0 | 1 | 0 | 1 |
| 0 | 1 | 1 | 0 |
| 1 | 0 | 0 | 1 |
| 1 | 0 | 1 | 1 |
| 1 | 1 | 0 | 0 |
| 1 | 1 | 1 | 0 |

POS Equation: $y = (a+/b+/c)(/a+/b+c)(/a+/b+/c)$

This next example truth table, shown below, yields a POS equation of: $y = (A+/B)$

| a | b | y |
|---|---|---|
| 0 | 0 | 1 |
| 0 | 1 | 0 |
| 1 | 0 | 1 |
| 1 | 1 | 1 |

## Programmable Logic Devices

Programmable logic devices (PLDs) are digital logic chips that can be programmed by the engineer or circuit designer. The PLD can store programs, data, lookup tables, digital logic equations, and state machines. One single PLD can replace many off-the-shelf digital logic chips. This is a tremendous advantage. Not only is there a reduction in cost, size, and weight, but also in power used. Reliability and speed increase.

The cost of a PLD is only about a few dollars. Types of PLDs include EPROM, EEPROM, PALs and PLAs, and GALs. If the engineer makes a mistake or an upgrade is needed, the PLD can be reprogrammed. PLDs are found in most electronic devices such as computers.

For example, to implement the function y1 = Not ((alarm1 OR alarm2) AND sensor1) would require three different chips. Or, this function could also be implemented using one GAL16v8. The GAL16v8 has up to 16 inputs and 8 outputs, whereas the GAL20v8 has up to 20 inputs and 8 outputs. The equation to program y1 into the PLD would be: y1 = !((alarm1 # alarm2)&sensor1). This example program shows just how easy it is to program a few simple equations into a PLD.

```
Name gates;
Designer T. Grace;
Device G20V8;
Date 8 sept 05;
// Inputs: Define inputs -----------------------
    Pin 1 = a;
    Pin 2 = b;
// Outputs: Define outputs --------------------
    Pin 17 =   y2; /* active high */
    Pin 18 = !y3; /* active low   */
// Logic: Define output equations --------------
    y2 = !b;
    y3 = a # b;       /* if a or b then y3 is low */
```

## Timer, LM555

This timer chip has three modes of operation: astable, monostable, and bistable. Those names refer to the output types generated. astable is a square wave with a specified duty cycle, monostable has one stable state (high or low), and bistable is stable in both high and low conditions.

**The astable mode.** How does one design a circuit in the astable state to produce a square wave? In this mode of operation, the output of the 555 is a square wave with a specified frequency and duty cycle. Frequency is equal to 1/period, while the duty cycle is time that the wave is high divided by the period. Here is the astable circuit with the pin numbers shown:

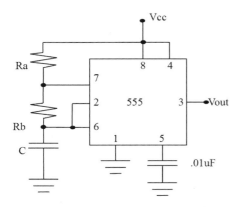

For any circuit:

freq = 1 / period

period = t high + t low

duty cycle = t high / (t high + t low)

For this circuit:

t high = .693 (Ra + Rb) * C

t low = .693 Rb * C

This example shows how to calculate Ra and Rb for the astable oscillator circuit given a 2 uf capacitor, a duty cycle of 70%, and a frequency of 20kHz.

T = 1/f = .00005 sec = t high + t low

0.70 = t high / (t high + t low)

Thus: t high = .000035 sec

and t low = .000015 sec

Then Rb = 10.8 ohms

and Ra = 14.4 ohms

## Multiplexer

A multiplexer is a chip that has several or many inputs (2, 4, 8, 16, 32, etc.) and one output channel. The multiplexer is used to select one of these inputs. The inputs are typically digital, but some multiplexers allow for analog input signals.

The select lines going into the chip determine which input is selected. One and only one of the inputs is selected by the select line to be connected to the output channel. For example, this chip would be useful if there was just one digital input available on the microprocessor but many digital values to read. The microprocessor would determine via the select lines which sensor to select. If there are three select lines, then there should be eight inputs, as $2^3 = 8$. Four select lines would allow for 16 inputs.

Some chips have an enable-input pin that enables or disables the output. Some multiplexers can accept only digital inputs, while others can accept analog signals.

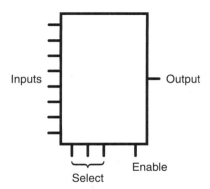

## Current and Voltage Limitations

The output of a gate or chip cannot source or sink infinite current. In fact, the current available from a typical 7400 series gate is quite low. The maximum output and input currents are:

$I_{OH}$ = maximum current when output is high. About −0.4 mA for TTL

$I_{OL}$ = maximum current when output is low. About 16.0 mA for TTL

$I_{IH}$ = maximum current when input is high. About 0.04 mA for TTL

$I_{IL}$ = maximum current when input is low. About −1.6 mA for TTL

If a low-resistance load is connected to the output of a gate, one might exceed these currents and thus the output voltage will deviate from the ideal. Instead of a high output being 5 volts, it could decrease to something lower. If the output goes low enough, the voltage will no longer be considered a high voltage.

A high is 5v while a low is 0v. These values are the ideal, but in reality, signals will vary from the ideal. If 4 volts is received by an input of a TTL gate, how is it perceived? As a high or a low? The answer to that is:

For signals going into a TTL gate:

Any voltage from 2 to 5v is seen as a high.

Any voltage from 0 to 0.8v is seen as a low.

So 4v will be interpreted as a high. But what if the voltage is 1.2v? This is in the uncertain region and how it is perceived is undetermined. It might be a high or it might be a low. One cannot tell. Obviously, then, we do not want our signals to be in this region.

As for an output from a gate, the ideal values are 5v and 0v. In practice, the voltages may be somewhere in between due to a load-drawing current. If a load draws too much current, then the 5 volts may drop down until it is no longer considered a high. The output specification for TTL gates is:

For signals coming out from a TTL gate:

A high is any voltage from 2.4 to 5v.

A low is any voltage from 0 to 0.4v.

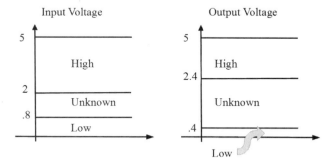

## Fan Out

The output of one chip may be connected to the input of another or even several other chips. Keep in mind that the output of a chip does not have an infinite amount of power but rather a finite amount of power. This means that one output may drive a limited number of inputs before the signal degrades. The number of gates that can be driven by one output is called *fan out*. This is determined by the output current and input current. The fan out may be different when driving a low versus driving a high.

Consider one gate driving many gates.

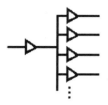

Notice that the direction of $I_{OL}$ is back into the output of the gate.

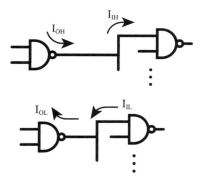

$I_{OH}$ = maximum current when output is high. About −0.40 mA for TTL

$I_{OL}$ = maximum current when output is low. About 16.00 mA for TTL

$I_{IH}$ = maximum current when input is high. About 0.04 mA for TTL

$I_{IL}$ = maximum current when input is low. About −1.60 mA for TTL

Fan out is the minimum of $(I_{OH}/I_{IH})$ and $(I_{OL}/I_{IL})$. For the numbers given, the fan out is 10. Again, this means that up to 10 gates can be driven by one gate before it becomes unreliable.

## Resistors, SIPs, and DIPs

Resistors can be packaged singly or in a group. The value of a single resistor is specified by three bands of color on the device. To read the value, refer to the color code for resistors. A very common resistor is the 1k ohm, with colors of brown, black, and red.

A DIP is a dual in-line-package that, in this case, contains multiple resistors. There are two pins per resistor. There is no power supply needed. The value of the resistance could, of course, be measured on a meter, but the package is also usually marked with a number. The third digit of the number represents the number of zeros after the first two digits. For example, a DIP marked 471 is 47 with one zero, or 470 ohms. A 102 is a 10 with two zeros after it, or 1000 ohms.

In this image of a DIP resistor, notice that the value "102" means 1000 ohms.

 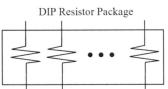

DIP Resistor Package

A SIP is a single-in-line-package that, in this case, contains multiple resistors. The number of resistors is one less than the number of pins. One pin is common to one end of all resistors. The value of the resistance could, of course, be measured on a meter, but the package is also usually marked with a number. The third digit of the number represents the number of zeros after the first two digits. For example, a SIP marked 102 is 10 with two zeros, or 1000 ohms. SIPs are well suited as pull up resistors.

In the following image of a SIP resistor, notice the marking "471" (meaning 470 ohms) and the dot in the picture. This resistor pack has one pin that is common to all resistors.

SIP Resistor Package

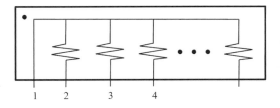

Dot. Pin 1 is Common

## Logic Switches

A switch can be used to turn something on, but it can also be used to select a high or low voltage, which is often sent to the input of a computer. This is one way for the user of the microprocessor to tell the microprocessor what to do. Following is a switch circuit that will produce 5v or 0v depending on the position of the switch. With the switch closed, the output is set to ground and is 0v. With the switch open, the output voltage is very nearly 5v.

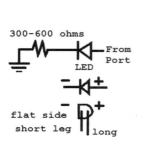

The value of the resistor, if too large, will cause a significant voltage drop from the 5v, and if there's too much of a drop, then the voltage will no longer be high. On the other hand, if the resistor is too small, when the switch is closed, the current will be larger than desired. Generally, 1k to 10k ohms works well.

## LEDs

An LED is a *light-emitting diode*. It lights up when current flows through it in the forward direction. Current can flow only in the forward direction to light the LED. Most round LEDs have a small flat side on the case. This side is the ground, or negative, side. The value of the resistor can be anywhere from 220 to 1000 ohms or thereabouts. Common resistors values are 220, 330, and 470 ohms. SIPs and DIPs work well when a bank of resistors are needed. If the LED does not light up in the circuit, try reversing the legs of the LED.

This diagram and image illustrate how to connect an LED and how to locate the + and − side of an LED. Some LEDs are round, while others may be square.

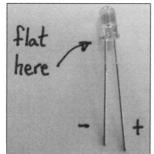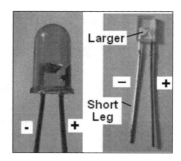

In this picture, the flat (negative) side is on the left. It's difficult to discern the flat side from this image, but you can determine the left side is negative by the larger metal area inside the LED, as well as the shorter leg.

LEDs also come packaged in groups or banks. This one contains 10 LEDs. These will still need current limiting resistors. One side is (+) and the other (−).

## Seven-Segment Displays

Seven-segment displays get their name from the fact that they have seven LED segments. These displays are used in machines such as older calculators, vending machines, home appliances, etc. Each display represents one digit of a number. A soda machine might use three of them to represent the amount of money entered.

These displays need seven wires going to them to control each individual segment. Also, most displays require a resistor for each segment in order to limit the current flow. Typical values range from 220 to 1000 ohms. Higher resistance would produce a dimmer display. Each segment in the display is produced by an LED. Some displays are common anode (common Vcc), while others are common cathode (common ground).

The common anode (Vcc) means that all of the LED's anodes are connected to power, and to turn on a segment, send 0v to that particular segment. These parts are low active:

TIL312

NTE3061

LSD3221

NTE3052

The common cathode (ground) means that on one side all LEDs are connected to ground, and to turn on a segment, send 5v to that particular segment. These parts are high active:

FND503

ECG3079

LSD3211

The pin-out for the seven-segment display is illustrated next. Notice the seven segments labeled A through F. There are many different seven-segment displays manufactured. For example, on the TIL312, pin 1 is connected to segment A. The package may not have all pins populated as shown in the picture. DP refers to the decimal point.

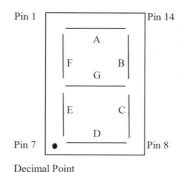

Following is a seven-segment display pin-out chart. The display pin numbers are located on the top row. The letters A through G refer to one of the seven segments that light up. Vcc and GND are power and ground, respectively.

| Pin # --> | 1 | 2 | 3 | 4 | 5 | 6 | 7 | 8 | 9 | 10 | 11 | 12 | 13 | 14 |
|---|---|---|---|---|---|---|---|---|---|---|---|---|---|---|
| Display: | | | | | | | | | | | | | | |
| TIL312 | A | F | Vcc | | | | E | D | DP | C | G | | B | Vcc |
| NTE3061 | A | F | Vcc | | | DP | E | D | | C | G | | B | Vcc |
| LSD3221 | A | F | Vcc | | | | E | D | DP | C | G | | B | Vcc |
| NTE3052 | A | F | Vcc | | | | E | D | DP | C | G | | B | Vcc |
| LSD3211 | F | G | | GND | | E | D | C | DP | | | GND | B | A |

Seven-segment displays are often driven by a 7447 chip. This chip takes a 4-bit base 2 number and determines which of the seven segments to light up. Its output is low active, which means it is well suited for the common anode displays.

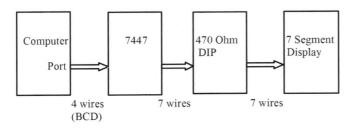

For example, if the 4-bit BCD number is 0110, a 6 should be seen, and thus segments A, F, G, C, D, E will light up. The 7447 chip needs power and the four binary bits. It generates the seven segments A through F. If the binary number is above 9, then the output is meaningless. A 470-ohm DIP is a handy way to limit the current through the seven-segment display.

The 7447 inputs a 4-bit BCD number on d3, d2, d1, and d0, with d0 being the LSB. Pins A through G output to the seven-segment display. Each of the seven segments of the display is controlled individually by one of these output lines. RB stands for register blanking and does not need to be connected.

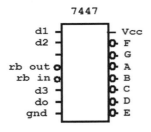

# CHAPTER 2

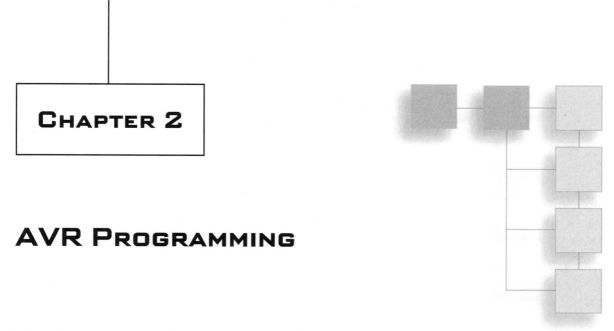

# AVR PROGRAMMING

This chapter shows you how to program the AVR first in assembly, then in C, and then in C++. Assembly has the advantage of speed, while C has the advantage of convenience. In some cases, C++ (using objects) can make for better code than C. First, though, you should understand the features, registers, and architecture of the AVR.

## MICROCOMPUTER ARCHITECTURE

This section describes in general microcomputer architecture and how a microcomputer works. A computer contains a microprocessor (MPU or CPU) that is connected to devices such as memory, drives, switches, lights, and so on, by wires or bus. The CPU runs the program and does the computing. This computing is done within the hardware of the CPU. The number of registers within the CPU varies from one CPU to another. More registers are better because it makes computing easier. Also, larger registers (more bits) make it easier to deal with larger numbers. The downside is that the CPU is larger, more costly, and uses more power. Why use a large expensive processor to run a toaster?

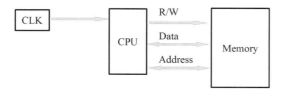

Regarding the AVR microprocessor, the SRAM is located on the microprocessor itself, and each SRAM memory location is 8 bits. When reading from or writing to a memory location, each location is 1 byte (8 bits). Thus, the smallest number is zero and the largest is 0xff or 255 or 0b11111111. The bus system is a collection of wires connected to the CPU that can be grouped into three areas: data, address, and control.

The data bus is the pathway for data to travel. On the AVR, it is 8 bits wide. The clock frequency of the computer determines the speed of operation (although a faster clock uses more power). The clock synchronizes the separate components of the computer. The frequency of many of the AVRs default to 1 MHz but can be made to go faster or slower.

The address bus determines what memory location the CPU will talk to. The address bus selects, or points to, the memory location that will be read from or written to. If a CPU has 16 address lines, it can address $2^{16}$ memory locations, or 65536 locations. The first address is 0x0000 and the last is 0xffff.

The control bus is a collection of signals that determine such things as which way the data will travel. For example, the R/W line is high when the MPU is reading memory and low when the MPU is writing to memory. Another control bus line is called RESET. When it is brought low, it will tell the CPU to reboot. The CPU also needs power (Vcc is usually 5v or 3.3v) and ground.

Computer memory can be thought of as a linear collection of empty boxes or perhaps houses on a street. Each box or house can hold a number. The maximum size of the number held in memory depends on the number of bits used for each memory location. If the memory is 8 bits wide ($2^8 = 256$), then the maximum value that can be stored is 255 (since values start from 0). If the memory is 16 bits wide ($2^{16} = 65536$), the maximum value is then 65535. A wider memory does make it easier to work with larger numbers. The next question might be, how many houses can we have on the street? This depends on the number of address lines. If we have four address lines, then $2^4$ allows for only 16 houses, each with a unique address. More address lines allow for more houses. But more lines add to the cost and complexity of a computer. With 12 address lines, we can have $2^{12}$, or 4096 unique addresses.

External hardware such as sensors or motors can be connected to the computer. The connections are sometimes done through ports. The port is often mapped into memory. This means that the port is a memory location. Reading or writing to that memory location is the same as communicating to what is connected to the port.

# THE AVR FAMILY OF MICROCONTROLLERS

ATMEL has a wide range of microcontrollers available for use in all sorts of electrical devices. The engineer needs to pick the microcontroller best suited for the task. For example, toasters, refrigerators, coffee makers, kids' toys, small robots, and vending machines do not need much memory or I/O from the processor. In this case, the ATtiny13, which is very cheap and small, might be a good choice. It is small, low power, inexpensive, and has a few I/O pins. On the other hand, some applications need more resources, such as machines or appliances with graphic or touch screens, or devices with many input buttons. For those applications, one might choose the ATmega328, which has more memory and more I/O. After programming the chip, it may be removed from the development board and placed in the application circuit. The application circuit will need to supply power.

Consider the electronics in a car. One could use a touch screen for the user interface to control the temperature, radio, phone, speed, engine, etc. Also, each wheel might have individual ABS braking. Instead of having one large processor trying to do everything, it would be better to have one small processor for each individual system. Thus, each wheel would have its own processor to control that wheel's ABS system. Using one processor to service all tasks needed by the car might not be fast enough, or the software might be too complex. Instead, these smaller processors can be distributed throughout the car to divide the work. They would be tied back to a central processor through an SPI or CAN bus. The CAN bus is used as the standard in the car industry.

The ATtiny13 and ATtiny85 have 8 pins, the ATmega328 has 28 pins, and the ATmega32 has 40 pins. Certainly, with more pins, you can have more input and output signals. There is very little difference in the code or software that runs on these chips. Before you start to program in assembly, you must first become familiar with the registers of the microprocessor (MPU). When programming, it is common to load a register (like r16) with a number, then perform a mathematical operation on the number, and then finally store it to memory or to a port. The inputs and outputs to external hardware are done via the ports. These ports are memory mapped, which means that a port is a location in memory with a specific address.

## AVR Features

The ATtiny13, ATtiny85, ATmega328, and the ATmega32 can come in a DIP through-hole package, which makes it convenient to breadboard. The AVR family of microcontrollers uses separate areas for data and program memory. Programs are stored in Flash memory, while registers and data are stored in SRAM. Most instructions execute in one cycle. Fully static operation is possible (the CPU clock may drop to zero to save energy).

For more information, refer to the data sheets found at www.atmel.com. On the website, select 8-bit microcontrollers, and then select the specific microcontroller. Under Documents, you can download the PDF data sheets.

ATmega32, ATmega328, and ATtiny13:

ATtiny13/ATtiny85/ATmega328 Development Board:

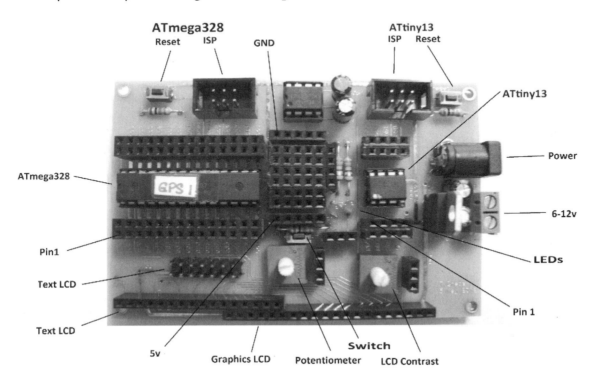

ATmega32 Development Board:

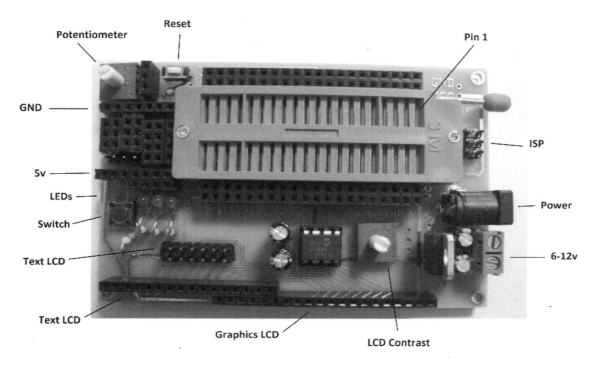

### *ATtiny13*

The ATtiny13 is one of the smallest and cheapest of the AVR microcontrollers. It takes up very little room on a circuit board, which saves money, especially in large-volume production. With very few pins and a small memory, it would not be well suited for such things as controlling a graphics LCD. Some of the product features and details are:

Cost: $2/ea.

Package: 8-pin DIP with 6 I/O lines

Four 10-bit A/D channels

Clock Frequency up to 20 MHz

SRAM Memory

       32 general-purpose registers, 0x00-0x1f

       64 I/O registers, 0x20-0x5f

       64 bytes for data, 0x60-0x9f

Flash Memory

> 1K for programs, 0x0000-0x03ff

> Arranged in 512 locations x 16 bits wide

EEPROM Memory

> 64 bytes, 0x00-0x3f

ATtiny13 Pin-Out:

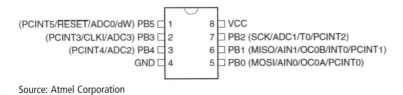

Source: Atmel Corporation

ATtiny13 and ATtiny85 Programmers' Model:

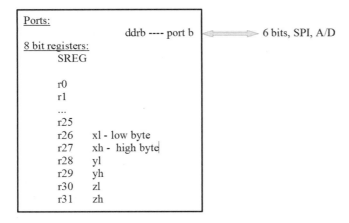

### *ATtiny85*

The '85 is very similar to and looks just like the '13 but contains more memory if one needs to store more data or larger programs. Some of the product features and details include:

Cost: $3/ea.

Same as the ATtiny13 but has:

> Flash Program - 8K

> SRAM - 512 bytes

EEPROM - 512 bytes

A/D has a programmable 20x gain.

### ATmega328

The '328 is the same chip used on the Arduino board. With 28 pins, it is not too big to breadboard nor too small and lacking in I/O (Goldilocks would have picked this chip). Some of the product features and details are:

Cost: $4.5/ea.

Instructions are essentially the same as for the ATtiny13, but there are more ports.

32K Flash Program

1K EEPROM

2K SRAM

Package: 28 pins with 23 I/O lines

Six 10-bit A/D channels

Six PWM channels

SPI, USART, Analog comparators

Clock Frequency up to 20 MHz.

ATmega328 Pin-Out:

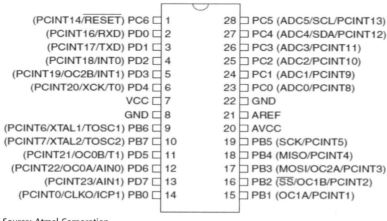

Source: Atmel Corporation

There are three ports on the '328: portb, portc, and portd. The direction of the port is determined by its DDR (DDRB, DDRC, DDRD), or *data direction register*. Each bit value and bit direction of the port is individually controllable. If a bit in the DDR is 0, then the corresponding bit in the port is an input, while a 1 in the DDR bit makes the port bit an output.

The 8-bit register SREG refers to the eight flags that can be used to make decisions. Each flag is 1 bit and is determined by the result of each instruction. Calculations are done in the general-purpose registers r0 to r31. The index registers are named x, y and z. Each is 16 bits, thus x is the concatenation of r27:r26.

ATmega328 Programmers' Model:

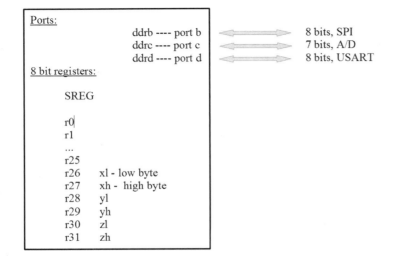

ATmega328, Detailed View:

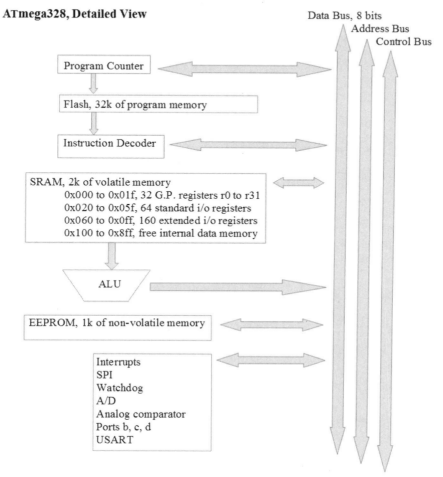

Source: Atmel Corporation

Notes:

Accessing Flash program memory requires instructions: lpm, spm

Accessing SRAM 0x00 to 0x1f one uses instructions: sbis, sbic, cbi, sbi

Accessing SRAM 0x00 to 0x3f one uses instructions: in, out

Accessing SRAM 0x60 to 0xff one uses these instructions: ld, lds, ldd, st, sts, std

## ATmega32

The '32 is a large 40-pin chip that has four ports for when one needs a lot of I/O. Some of the product features and details include:

Cost: $9/ea. (with 40 pins and 32 I/O lines)

32K Program space

1K EEPROM

2K SRAM

JTAG, SPI, USART, Analog comparators, 4 PWM channels

Eight 10-bit A/D channels with programmable gain.

Clock Frequency up to 16 MHz.

ATmega32 Pin-Out:

```
   (XCK/T0) PB0 ⊏ 1      40 ⊐ PA0 (ADC0)
       (T1) PB1 ⊏ 2      39 ⊐ PA1 (ADC1)
 (INT2/AIN0) PB2 ⊏ 3     38 ⊐ PA2 (ADC2)
  (OC0/AIN1) PB3 ⊏ 4     37 ⊐ PA3 (ADC3)
       (SS) PB4 ⊏ 5      36 ⊐ PA4 (ADC4)
     (MOSI) PB5 ⊏ 6      35 ⊐ PA5 (ADC5)
     (MISO) PB6 ⊏ 7      34 ⊐ PA6 (ADC6)
      (SCK) PB7 ⊏ 8      33 ⊐ PA7 (ADC7)
           RESET ⊏ 9     32 ⊐ AREF
             VCC ⊏ 10    31 ⊐ GND
             GND ⊏ 11    30 ⊐ AVCC
           XTAL2 ⊏ 12    29 ⊐ PC7 (TOSC2)
           XTAL1 ⊏ 13    28 ⊐ PC6 (TOSC1)
      (RXD) PD0 ⊏ 14     27 ⊐ PC5 (TDI)
      (TXD) PD1 ⊏ 15     26 ⊐ PC4 (TDO)
     (INT0) PD2 ⊏ 16     25 ⊐ PC3 (TMS)
     (INT1) PD3 ⊏ 17     24 ⊐ PC2 (TCK)
     (OC1B) PD4 ⊏ 18     23 ⊐ PC1 (SDA)
     (OC1A) PD5 ⊏ 19     22 ⊐ PC0 (SCL)
     (ICP1) PD6 ⊏ 20     21 ⊐ PD7 (OC2)
```

Source: Atmel Corporation

ATmega32 Programmers' Model:

```
Ports:                ddra ---- port a    <=====>    8 bits, A/D
                      ddrb ---- port b    <=====>    8 bits, SPI
                      ddrc ---- port c    <=====>    7 bits
                      ddrd ---- port d    <=====>    8 bits, USART
8 bit registers:
      SREG

      r0
      r1
      ...
      r25
      r26    xl - low byte
      r27    xh - high byte
      r28    yl
      r29    yh
      r30    zl
      r31    zh
```

## AVR CPU Registers

The AVR has numerous registers that allow the programmer to do calculations, move data, access memory and I/O, and to make decisions. There are more registers on the AVR than listed in the following section. These are the most common registers used in programming.

### General Purpose Registers

There are 32 general-purpose registers available for use to do addition, subtraction, or other mathematical operations. These 32 one-byte registers are mapped into the first 32 bytes of SRAM (the data memory). Only r16 to r31 can use immediate addressing. An example of immediate addressing is: ldi r16,0xfe which will load immediate r16 with the hex number fe. The registers r26 to r31 can be used for indexed addressing. Indexed addressing is a method of accessing memory by specifying the memory address to access. It is useful for when you want to access a range of sequential memory locations.

| Register | Memory Location in SRAM |
|----------|-------------------------|
| r0       | 0x00                    |
| r1       | 0x01                    |
| r2       | 0x02                    |
|          | *(Continued)*           |

| Register | Memory Location in SRAM |
|----------|-------------------------|
| ...      |                         |
| r26      | 0x1A, X-register Low Byte |
| r27      | 0x1B, X-register High Byte |
| r28      | 0x1C, Y-register Low Byte |
| r29      | 0x1D, Y-register High Byte |
| r30      | 0x1E, Z-register Low Byte |
| r31      | 0x1F, Z-register High Byte |

### Index Registers

Index registers are used to access data in a list or in sequence. Examples of when to use index addressing would be adding up a list of numbers, searching for something in a list of numbers, or sorting a list of data.

Three index registers (X, Y, and Z) are used in index addressing. They are 16 bits each. The index registers X, Y, and Z are mapped to registers r26 to r31 as shown in the preceding table. With 16 bits, each register can address up to 65536 different locations.

### Program Counter

Programs are stored in Flash memory. As the program is executed, the *program counter* (PC) points to the next instruction in the program to execute. The PC is a 14-bit wide register. Think of the PC as a pointer to a Flash memory location or the address of each instruction to execute. Each Flash memory location on the AVR is 2 bytes wide. Since the PC is 14 bits (same width as the address bus) it can point to 16K ($2^{14} = 16 \times 1024$) different memory locations. The programmer does not need to access or alter this register directly, as the AVR automatically updates it as the program runs.

As the first instruction is executed, the PC will point to the second instruction. When the second instruction is executed, the PC points to the third instruction, and so on. The PC is the MPU's mechanism for keeping track of where the next instruction is. An org statement determines the location in Flash memory where the program is to be stored. The org statement is not an instruction, nor does it initialize the PC, but rather tells the compiler where to place the program in memory. When power is applied or when reset occurs, the PC is set to 0x0000. This is where a program should start. The org directive looks like this:

```
.org    0x0000
```

### Status Register

The status register, SREG, is made up of eight flags. Each flag is a bit. The flags are used to make decisions such as whether a result is negative, positive, or zero. Another important decision to answer is whether a result exceeds the limits of signed or unsigned numbers. When a result exceeds the signed number limit, the V flag is set. When the unsigned number limit is exceeded, the C flag is set. So, which flag is used to test whether or not a result is zero? That would be the zero flag. The following table lists the flags and their meanings. The last four are the most often used flags.

| Flag | Name | Meaning |
|------|------|---------|
| I | Global Interrupt Enable | 1 = Allow interrupts. Use SEI and CLI |
| T | Bit Copy Storage | Use with BLD and BST instructions |
| H | Half carry | Set when there is a carry from bit 3 to bit 4 |
| S | Sign Bit | S = V exclusively ORed with N |
| V | Overflow | Set when result exceeds the range of signed numbers |
| N | Negative | Set when result is negative |
| Z | Zero | Set when the result is zero |
| C | Carry | Set when the result exceeds the range of unsigned numbers |

### Stack Pointer

The stack is an area of memory used to store data temporarily. It is very similar to a stack of plates on a shelf. Plates are pushed onto the stack and pulled off. The last plate pushed in is the first plate to be pulled off (LIFO). One cannot get to the bottom plate without first removing all of the plates above it. We cannot store an infinite number of plates because at some point we would hit the ceiling. This is very similar to what is occurring in the stack memory.

The stack is located in the SRAM memory. The *stack pointer* (SP) points to the stack data in the SRAM. It is initialized to a high memory location. SP points to the next available memory location and moves toward lower memory each time data is stored. The stack is also used in order to return from a subroutine. A call to a subroutine stores the return address on the stack. Typically, you do not need to manipulate the SP or the program counter, as they are automatically adjusted by the AVR. This relieves the programmer from that tedious task.

## AVR DEVELOPMENT SYSTEM

To program the AVR you will need to acquire these items:

- AVR Studio from Atmel
- An AVR development board with an ISP programming channel
- ISP programming cable
- An AVR such as the ATtiny13, ATtiny85, ATmega328P, or ATmega32

AVR Studio is a free download from Atmel. Version 6 works well and is used herein. Without any hardware, you can use it to simulate and debug programs, which is very handy. The AVR chips are relatively inexpensive. Through-hole DIP packages are much easier to work with than surface-mount AVRs, in addition, if they are socketed, they can easily be removed. Look for an ISP programming cable that has six pins (3 × 2) such as the USB AVR Programmer (part #1300) from Pololu.

## AVR Development Board

The development board can be made or purchased. Boards with a socketed AVR allow the AVR to be replaced if damaged. In addition, the AVR can be removed from the development board and placed onto a smaller board specially made for and dedicated to a project. This approach is flexible and saves money. A surface-mount AVR, for the most part, is not removable from the board. Here is a picture of a development board.

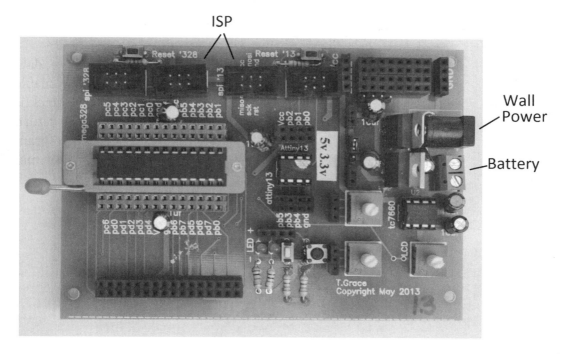

## Buy a Development Board

Look for a development board compatible with AVR Studio and an ISP programming cable. These components and AVR chips can be found from Pololu, SparkFun, Hobby Engineering, Grace Instrumentation (www.graceinstrumentation.com), and other suppliers. Or, instead of buying a development board, consider making your own. The simplest way is to use a protoboard and hand wire it. A more elegant method is to design your own printed circuit board (PCB).

## Use a Protoboard

Using a protoboard is the cheapest, quickest, and not exceptionally difficult method of programming the AVR for use (although it is not as neat and reliable as making a PCB). You will need an AVR, a protoboard, a 5v power supply, an ISP programming cable, and AVR Studio. A voltmeter will be useful for testing voltages.

Connect the AVR as shown in the following diagram and picture. More information for this connection can be found by searching the Internet. Be sure to get the ISP lines correct. You can use an LM7805 voltage regulator circuit to get 5v as shown by this circuit. Once this is done, proceed to the section "AVR Studio."

| ISP Pin | Signal | '13/'85 Pin | '328 Pin | '32 Pin |
|---------|--------|-------------|----------|---------|
| 1 | MISO | 6 | 18 | 7 |
| 2 | Vcc | 8 | 7 | 10 |
| 3 | SCK | 7 | 19 | 8 |
| 4 | MOSI | 5 | 17 | 6 |
| 5 | Reset | 1 | 1 | 9 |
| 6 | GND | 4 | 8 | 11 |

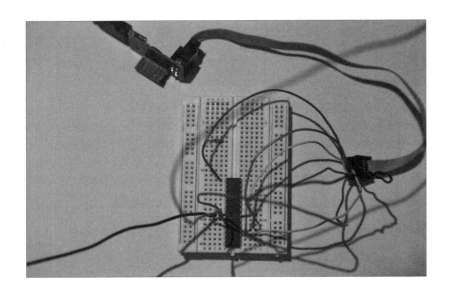

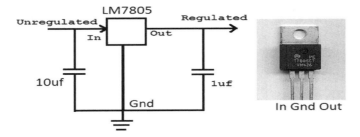

### Make a PCB Development Board

Making a printed circuit board (PCB) is slightly more expensive and time consuming but very rewarding. Be very careful before submitting the work to be manufactured. It is not uncommon to have mistakes in the design, resulting in having to redo the work. It might be a good idea to try the design on a protoboard first.

The steps are as follows.

1. Decide on the circuitry desired.

2. Download PCB software such as ExpressPCB or P2P to lay out the circuit.

3. Double-check your work before ordering.

4. Assemble the board and parts.

Some things to consider include:

One side of the board is for components, and the other is the solder side.

Leave enough space on the board for the part to fit without touching other parts.

Make the hole sizes large enough so that the pins go through. Not all parts have the same pin size. It is best to acquire the parts before designing the board.

Orient all the chips in the same direction.

Signal traces are generally .01 inches wide. Power lines are generally .04 inches.

A ZIF socket is handy if the AVR is removed frequently. Leave adequate room around the ZIF sockets since they are larger than regular DIP sockets.

Don't forget to include mounting holes in your design.

Items you may want to include on a board include:

Female headers to provide access points.

A Reset button.

A 3.3v regulator, jumper selectable.

Potentiometers, LEDs, switches.

A keyed or one-way ISP housing.

A 9v battery connection and a power plug connection.

An adequate number of decoupling capacitors.

The next three pictures show a minimal development board with an ATmega328, ATtiny13, ISP connections, and a 5v and 3.3v regulator. There are two power connectors. One is for a wall plug and the other is for a battery. The outputs of the regulators go to a jumper that will select 5v or 3.3v. S in the picture is for a reset switch. The 1k is the reset pull-up resistor. Notice the orientation of the signal lines on the ISP connector.

Here's the development board, showing all layers, the top layer, and the bottom layer.

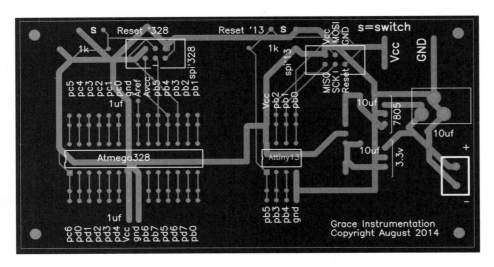

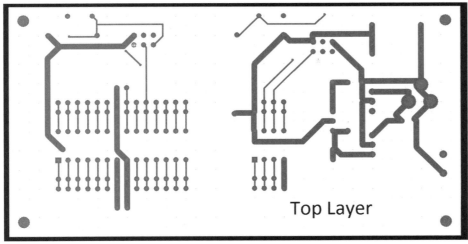

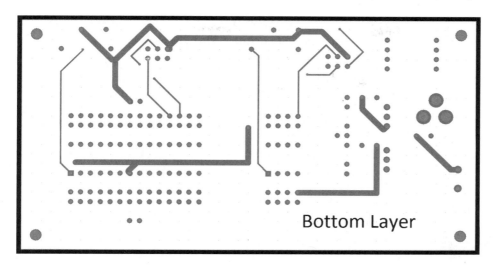

Bottom Layer

## AVR Studio

AVR Studio will be used to write programs and download code to the target AVR. It can be downloaded from Atmel for free. To download to an AVR, you will need a development board, which was covered in the prior section.

To start a new project in AVR Studio (Version 6), follow these steps:

1. Run AVR Studio and select File > New > Project.
2. Click on AVR Assembler (for assembly) or C and C Executable Project (for C programming). Enter a name, such as ATmega328_program1 and a file location. Click OK.
3. Now select a device from the list, such as the ATmega328p or type ATmega328p into the search box in the upper right. Select ATmega328p and click OK.
4. Type in the first program, as shown later in this chapter. (Program Test1)
5. To compile it, click Build > Build solution.

The simulator can be used to see how a program runs without having an actual AVR to run on. The simulator is also nice to use if you want to know how long a section of code takes to run.

- If the program built with no errors, then you can try to run it line by line on the simulator by selecting Debug > Step Into (F11). AVR Studio may ask you to select the debugging tool, which will be the AVR Simulator.

- In your program, a yellow arrow will point to the next instruction to execute. Bring up a window, via the icons at the top of AVR Studio, showing the values of the memory, processor, and registers. Each time you press F11 you will see how the registers, memory, and flags change as you move through the program.

- For a good resource on AVR programming, select Help > View Help. To stop debugging, select Debug > Stop Debugging.

To run a program on a development board, connect the board to the computer using the ISP programming cable and power the board.

- If the program built with no errors, then you can try to run it on the board by selecting Tools > AVR Programming or Device Programming. Select the STK500 and the correct com port as the tool. Note that you may need to first add the STK500 as a tool (which is done via Tools > Add STK500). Next, select the appropriate device that matches the chip you are programming. Click Apply. Optionally, you can read the device ID or the target voltage.

- Click the Memories option. Double-check to make sure that your program in hex format is selected as the program to download. Click on Program. This will download the program to the target and verify that it has worked. The program will run immediately on the target AVR.

- The program is now burned onto the chip and will reside there even without power. Anytime power or reset is applied, the program will restart.

AVR Studio 6 defaults to level 1 optimization when compiling C code. Level 1 tries to optimize code size and code speed. This means that code that has no apparent effect is removed. For example, a loop meant to delay for a millisecond is seen as doing nothing and is removed, which in turn could have a dilatory effect. To prevent this, set the optimization level to 0 by selecting Project > Properties > Tool Chain > AVR C Compiler > Optimization, and then set it to None -o0.

## PROGRAMMING IN ASSEMBLY

Programs in assembly are smaller and faster than programs in C or C++. However, the coding can sometimes be more tedious compared to C programming, especially in the case of multiplying, dividing, trigonometric functions, arrays, and so on. Programs in assembly are closer to the hardware and registers and are thus more efficient. A beginner programmer can start either with assembly or with C. Over time, many programmers migrate to C but revert to assembly when efficiency is needed. In either case, you will need AVR Studio and a development board as covered in the prior sections.

# A First Program in Assembly

Nearly every microprocessor has instructions to do the following operations. In order to be able to write programs, you must be familiar with these common instructions. Following is a list of some of the common assembly instructions, but it is not all-inclusive. For a complete list of instructions, refer to the instruction set. In most cases, the result of an operation is found in the leftmost operand. For example, add r2,r3 places the result of r2 plus r3 into register r2. Each program instruction takes up 2 bytes and is stored in a single Flash memory location. Flash memory starts at location 0x0000 and continues. The registers are numbered r0 to r31. You can load immediate into r16 to r31 but not r0 to r15. This means you can write ldi r16,2 but not ldi r15,2. Binary numbers are preceded with a 0b, and hex numbers are preceded with a 0x, while decimal numbers are the default and are not preceded with anything as shown by these lines:

```
ldi r16,0b0110
ldi r16,0xff
ldi r16,255
```

Some of the more common instructions are shown in the following table.

| Operation | Meaning | Instruction | Operand |
|---|---|---|---|
| add | Add register to register. Here r0=r0+r1 | add | r0,r1 |
| and | And the bits of a register with a number. | andi | r16,0b01111 |
| clear | Loads register with zero. r3 = 0 | clr | r3 |
| compare | Compare a register with a number. | cpi | r17,221 |
| complement | Toggle each bit of register. I's complement. | com | r4 |
| decrement | Decrease register by 1. r6 = r6 - 1 | dec | r6 |
| increment | Increase register by 1. r7 = r7 + 1 | inc | r7 |
| load | Load register with a number. r18 = 4 | ldi | r18,4 |
| move | Move register to register. r8 = r9 | mov | r8,r9 |
| multiply | Multiply number into register. | mul | r12,r13 |
| negate | Two's complement, or multiplying by -1. | neg | r11 |
| or | Or the bits of a register with a number. | ori | r19,0b11 |
| shift/rotate | Shift bits of register left or right. | ror | r10 |
| subtract | Subtract number from register. r20=r20-5 | subi | r20,5 |

## Program Test1

This program demonstrates how to blink two LEDs on port b that are connected to bits 0 & 1. The blinking will be too fast to see, as most instructions are 1 cycle, which is 1 microsecond at a frequency of 1 MHz. Comments begin with // or a semicolon. The org states where the program will be placed in memory. A hex number starts with 0x; a binary starts with 0b. Labels are names ending with a colon and are generally placed at the far left.

```
 ; ATmega328_test1.asm
.org      0x0000              ; Store in memory starting here.
    ldi    r16,0x0f           ; Load r16 with 0f hex.
    out    ddrb,r16           ; Make upper 4 bits of portb input.
                              ; Make lower 4 bits of portb output.
    ldi    r17,0x00           ; Load r17 with 00 hex.
    ldi    r18,0b11           ; Load r18 with 11 binary.
top:
    out    portb,r18          ; Turn on lowest 2 bits of port b.
    out    portb,r17          ; Turn off all bits of port b.

    rjmp   top                ; Jump to top.
```

For the preceding program, please note what each of these lines do.

```
ldi r16,0x0f       ; Loads a register, r16, with the number 0x0f.
out ddrb,r16       ; Sets the direction of port b according to the r16 value.
out portb,r18      ; Sends to port b the value in r18.
```

## Program Test2

This program shows how to blink an LED connected to port b bit 0, but with a delay subroutine added. The instruction rcall will call a subroutine. The instruction sbi will set a bit of a register, while cbi clears a bit. The DDRB is used to set the data direction of port b. One must set the direction of port b bit 0 to output by placing a 1 in the DDRB bit position 0. The #include tells the compiler to include the code found in the specified file.

```
 ; ATmega328_test2.asm
.org          0x0000
    ldi     r16,0b00001111   ; Make the lower 4 bits output
    out     ddrb,r16         ;     for port b.
top:
    sbi     portb,0          ; Set bit 0 immediate of port b
    rcall   delay_100ms      ; Calling a subroutine.

    cbi     portb,0          ; Clear bit 0 immediate of port b
    rcall   delay_100ms

    rjmp    top              ; Relative jump to label top
#include "i:\delays_1mhz.asm"   ; This file has the delay subroutine.
```

The next bit of code shows how to make a delay subroutine. The subroutine starts where the name or label of the subroutine is located. The label has a colon after it. For visual effect and readability, it is nice to have a horizontal line denoting the beginning and end of the subroutine. Notice that every subroutine ends with the instruction ret.

```
;delays_1mhz.asm
; -------------------------
delay_5ms:
      rcall    delay_1ms
      rcall    delay_1ms
      rcall    delay_1ms
      rcall    delay_1ms
      rcall    delay_1ms
      ret
; -------------------------
; -------------------------
delay_1ms:
      push   r16       ;save the value in r16
      ldi    r16,99    ;accounts for overhead of 12 cycles.
delay_1ms1:            ; 10 us/loop
      nop              ; [1 cycle ]
      nop
      nop
      nop
      nop
      nop
      nop
      dec    r16
      brne   delay_1ms1  ; [2 cycles ]
      pop    r16       ;restore the value in r16
      ret
; -------------------------------------
```

### Program Test3

The next program shows examples of adding, subtracting, subroutines, looping, arithmetic, and logic instructions. Note that the immediate instructions ldi, sbi, subi, and andi cannot be used with registers r0 to r15.

```
//  ATmega328_test3.asm
.org  0x0000
    ldi    r16,0x12          ;load immediate r16 with hex # 12
    ldi    r17,22            ;load immediate r17 with decimal # 22
    ldi    r18,0b00001111    ;load immediate r18 with binary # 1111
```

```
    inc   r0                ;increment r0
    dec   r1                ;decrement r1

    add   r2,r1             ;r2 = r2 + r1
    sub   r13,r17           ;r13 = r13 - r17

    subi  r22,11
;   subi  r2,11             ;Error. Cannot use subi or ldi on r0 to r15
;   addi  r22,11            ;Error. addi does not exist.

    rcall delay             ;Call a subroutine.
main0:
    rjmp  main0             ;Jump to main0

; ------------------------------ A subroutine
delay:
    nop           ; no operation
    ret           ; return from subroutine.
; ------------------------------
```

## Program Test4

Program 4 demonstrates how to read port b bit 1 as input and send it to port b bit 0 as output. Some of the bits of a port can be set to input while the other bits are set to output. When reading port b as input, the name of the port is pinb not portb. It is very common as a beginner to forget this name change even though it is essentially the same port. The instruction asr r16 shifts all the bits of a register to the right by one position. Notice that there was a nop in line 4 right after the out ddrb,r16. The reason for this is that it takes a small amount of time for the out ddrb,r16 to take effect. Without the nop, the next instruction (in r0,pinb) may not have the direction of the port set up in time.

```
; ATmega328_test4.asm
; This program shows input and output of a port.
.org    0x0000
    ldi   r16,0b01     ; Make portb bit 0 output, and
    out   ddrb,r16     ; make portb bit 1 input.
    nop                ; Allow time for ddrb to change.
more:
    in    r0,pinb      ; Read port b as input.
    mov   r16,r0       ; Transfer r0 into r16.
    andi  r16,0b10     ; Isolate bit 1, zero other bits.

;   andi  r0,0b10      ; This would be an error.
                       ; andi cannot use r0 to r15.

    asr   r16          ; Arithmetic shift right.
    out   portb,r16    ; Output r16 to portb.
    rjmp  more
```

### Program Test5

You should be familiar with the instructions in this program. Refer to the instruction set for more information on each instruction.

```
; ATmega328_test5.asm
.org   0x0000
   clr   r16        ; clear all bits
   ser   r17        ; sets all bits
   com   r18        ; one's complement
   neg   r19        ; two's complement
   or    r0,r1      ; or bits into r0
   and   r0,r1      ; and bits into r0
   inc   r3         ; increment r3
   dec   r4         ; decrement r4
   add   r3,r4      ; add r4 into r3
   sub   r3,r4      ; subtract r4 out of r3
   sbi   portb,3    ; sets bit 3 of port. 0000 1000
   cbi   portb,4    ; clears bit 4 of port. 0001 0000
   lsl   r0         ; shift  left
   lsr   r0         ;        right
   rol   r0         ; rotate left
   ror   r0         ;        right
more:
   rjmp  more
```

## AVR Studio Assembler Notes

For more help on AVR Studio, select Help > View Help when running AVR Studio. The following shows how to use different number bases when writing code.

```
   ldi   r16,0xff          ; 0x denotes a hex number
   ldi   r16,255           ; default is base 10
   ldi   r16,0b11110000    ; 0b denotes a binary number
```

The directive .org followed by a number determines where in memory the following instructions are placed. Typically, we start programs at memory location 0 because when power is applied the AVR starts code execution at memory location 0. The directive .equ means equate, and it replaces all occurrences of a name with a number. Using a name is informative because it has meaning versus using a number, which can have the same effect but conveys no meaning. In this program, an LED is on bit 5. In addition, when using an equate one can make a single change to the equate versus making many changes throughout the program. The directive db will define or store the following bytes in memory. Labels typically go at the leftmost column and are followed by a colon. The label is also equal to the address of the first byte of the data or instructions that follow. The #include will include the specified file. Comments begin with a semicolon.

```
.equ    led = 5             ; an equate
.org    0x0000              ; program starts at memory 00000

top:
   nop
   sbi    portb,led         ; set bit 5
   rjmp   top
msg1:    .db    "Hello mom",10,13
#include "i:\delays_1mhz.asm"    ; This file has the delay subroutine.
```

## Complete Instruction Set

Following is a complete list of the assembly instructions. The instructions work on most chips in the AVR family. Please note the important subtleties of these common instructions.

■ To move data between the general-purpose registers, r0 to r31, use mov.

■ To move data in or out of the I/O registers, use instructions in or out.

■ The immediate instructions like ldi work with registers r16 to r31 but not with r0 to r15.

■ Use the instructions add or sub to add or subtract between the general-purpose registers.

■ Note that there is a subi but no addi instruction.

| Instr. | Operand | Name | Meaning | Flags | Cycles |
|---|---|---|---|---|---|
| **Arithmetic Instructions (ATmega, not ATtiny)** | | | | | |
| MUL | Rd, Rr | Multiply Unsigned | R1:R0 ← Rd x Rr | Z,C | 2 |
| MULS | Rd, Rr | Multiply Signed | R1:R0 ← Rd x Rr | Z,C | 2 |
| MULSU | Rd, Rr | Multiply Signed with Unsigned | R1:R0 ← Rd x Rr | Z,C | 2 |
| FMUL | Rd, Rr | Fractional Multiply Unsigned | R1:R0 ← (Rd x Rr) << 1 | Z,C | 2 |
| FMULS | Rd, Rr | Fractional Multiply Signed | R1:R0 ← (Rd x Rr) << 1 | Z,C | 2 |
| FMULSU | Rd, Rr | Fractional Multiply Signed with Unsigned | R1:R0 ← (Rd x Rr) << 1 | Z,C | 2 |

## Arithmetic and Logic Instructions

| | | | | | |
|---|---|---|---|---|---|
| INC | Rd | Increment | Rd ← Rd + 1 | Z,N,V | 1 |
| DEC | Rd | Decrement | Rd ← Rd − 1 | Z,N,V | 1 |
| ADD | Rd, Rr | Add two Registers | Rd ← Rd + Rr | Z,C,N,V,H | 1 |
| ADC | Rd, Rr | Add with Carry two Registers | Rd ← Rd + Rr + C | Z,C,N,V,H | 1 |
| ADIW | Rdl,K | Add Immediate to Word | Rdh:Rdl ← Rdh:Rdl + K | Z,C,N,V,S | 2 |
| SUB | Rd, Rr | Subtract two Registers | Rd ← Rd - Rr | Z,C,N,V,H | 1 |
| SUBI | Rd, K | Subtract Constant from Register | Rd ← Rd - K | Z,C,N,V,H | 1 |
| SBC | Rd, Rr | Subtract with Carry two Registers | Rd ← Rd - Rr - C | Z,C,N,V,H | 1 |
| SBCI | Rd, K | Subtract with Carry Constant from Reg. | Rd ← Rd - K - C | Z,C,N,V,H | 1 |
| SBIW | Rdl,K | Subtract Immediate from Word | Rdh:Rdl ← Rdh:Rdl - K | Z,C,N,V,S | 2 |
| COM | Rd | One's Complement | Rd ← 0xFF − Rd | Z,C,N,V | 1 |
| NEG | Rd | Two's Complement | Rd ← 0x00 − Rd | Z,C,N,V,H | 1 |
| AND | Rd, Rr | Logical AND Registers | Rd ← Rd • Rr | Z,N,V | 1 |
| ANDI | Rd, K | Logical AND Register and Constant | Rd ← Rd • K | Z,N,V | 1 |
| OR | Rd, Rr | Logical OR Registers | Rd ← Rd v Rr | Z,N,V | 1 |
| ORI | Rd, K | Logical OR Register and Constant | Rd ← Rd v K | Z,N,V | 1 |
| EOR | Rd, Rr | Exclusive OR Registers | Rd ← Rd ⊕ Rr | Z,N,V | 1 |
| CLR | Rd | Clear Register | Rd ← Rd ⊕ Rd | Z,N,V | 1 |
| SER | Rd | Set Register | Rd ← 0xFF | None | 1 |
| SBR | Rd,K | Set Bit(s) in Register | Rd ← Rd v K | Z,N,V | 1 |

*(Continued)*

| Instr. | Operand | Name | Meaning | Flags | Cycles |
|--------|---------|------|---------|-------|--------|
| CBR | Rd,K | Clear Bit(s) in Register | Rd ← Rd • (0xFF - K) | Z,N,V | 1 |
| TST | Rd | Test for Zero or Minus | Rd ← Rd • Rd | Z,N,V | 1 |
| **Branch Instructions** | | | | | |
| RJMP | k | Relative Jump | PC ← PC + k + 1 | None | 2 |
| IJMP | | Indirect Jump to (Z) | PC ← Z | None | 2 |
| CPSE | Rd,Rr | Compare, Skip if Equal | if (Rd = Rr) PC ← PC + 2 or 3 | None | 1/2/3 |
| CP | Rd,Rr | Compare | Rd – Rr | Z, N,V,C,H | 1 |
| CPC | Rd,Rr | Compare with Carry | Rd – Rr – C | Z, N,V,C,H | 1 |
| CPI | Rd,K | Compare Register with Immediate | Rd – K | Z, N,V,C,H | 1 |
| SBRC | Rr, b | Skip if Bit in Reg. Cleared | if (Rr(b)=0) PC ← PC + 2 or 3 | None | 1/2/3 |
| SBRS | Rr, b | Skip if Bit in Reg. is Set | if (Rr(b)=1) PC ← PC + 2 or 3 | None | 1/2/3 |
| SBIC | P, b | Skip if Bit in I/O Reg. Cleared | if (P(b)=0) PC ← PC + 2 or 3 | None | 1/2/3 |
| SBIS | P, b | Skip if Bit in I/O Reg. is Set | if (P(b)=1) PC ← PC + 2 or 3 | None | 1/2/3 |
| BRBS | s, k | Branch if Status Flag Set | if (SREG(s)=1) then PC ← PC+k+1 | None | 1/2 |
| BRBC | s, k | Branch if Status Flag Cleared | if (SREG(s)=0) then PC ← PC+k+1 | None | 1/2 |
| BREQ | k | Branch if Equal | if (Z = 1) then PC ← PC + k + 1 | None | 1/2 |
| BRNE | k | Branch if Not Equal | if (Z = 0) then PC ← PC + k + 1 | None | 1/2 |
| BRCS | k | Branch if Carry Set | if (C = 1) then PC ← PC + k + 1 | None | 1/2 |
| BRCC | k | Branch if Carry Cleared | if (C = 0) then PC ← PC + k + 1 | None | 1/2 |

| BRSH | k | Branch if Same or Higher | if (C = 0) then PC ← PC + k + 1 | None | 1/2 |
|------|---|--------------------------|----------------------------------|------|-----|
| BRLO | k | Branch if Lower | if (C = 1) then PC ← PC + k + 1 | None | 1/2 |
| BRMI | k | Branch if Minus | if (N = 1) then PC ← PC + k + 1 | None | 1/2 |
| BRPL | k | Branch if Plus | if (N = 0) then PC ← PC + k + 1 | None | 1/2 |
| BRGE | k | Branch if Greater or Equal, Signed | if (N $\oplus$ V= 0) then PC ← PC+k+1 | None | 1/2 |
| BRLT | k | Branch if Less Than Zero, Signed | if (N $\oplus$ V= 1) then PC ← PC+k+1 | None | 1/2 |
| BRHS | k | Branch if Half Carry Flag Set | if (H = 1) then PC ← PC + k + 1 | None | 1/2 |
| BRHC | k | Branch if Half Carry Flag Cleared | if (H = 0) then PC ← PC + k + 1 | None | 1/2 |
| BRTS | k | Branch if T Flag Set | if (T = 1) then PC ← PC + k + 1 | None | 1/2 |
| BRTC | k | Branch if T Flag Cleared | if (T = 0) then PC ← PC + k + 1 | None | 1/2 |
| BRVS | k | Branch if Overflow Flag is Set | if (V = 1) then PC ← PC + k + 1 | None | 1/2 |
| BRVC | k | Branch if Overflow Flag is Cleared | if (V = 0) then PC ← PC + k + 1 | None | 1/2 |
| BRIE | k | Branch if Interrupt Enabled | if ( I = 1) then PC ← PC + k + 1 | None | 1/2 |
| BRID | k | Branch if Interrupt Disabled | if ( I = 0) then PC ← PC + k + 1 | None | 1/2 |

**Subroutine Instructions**

| RCALL | k | Relative Subroutine Call | PC ← PC + k + 1 | None | 3 |
|-------|---|--------------------------|-----------------|------|---|
| ICALL |   | Indirect Call to (Z) | PC ← Z | None | 3 |
| RET |   | Subroutine Return | PC ← STACK | None | 4 |
| RETI |   | Interrupt Return | PC ← STACK | I | 4 |

*(Continued)*

| Instr. | Operand | Name | Meaning | Flags | Cycles |
|---|---|---|---|---|---|
| **Bit Instructions** | | | | | |
| SBI | P,b | Set Bit in I/O Register I/O | (P,b) ← 1 | None | 2 |
| CBI | P,b | Clear Bit in I/O Register I/O | (P,b) ← 0 | None | 2 |
| LSL | Rd | Logical Shift Left | Rd(n+1) ← Rd(n), Rd(0) ← 0 | Z,C,N,V | 1 |
| LSR | Rd | Logical Shift Right | Rd(n) ← Rd(n+1), Rd(7) ← 0 | Z,C,N,V | 1 |
| ROL | Rd | Rotate Left Through Carry | Rd(0) ← C, Rd(n+1) ← Rd(n), C ← Rd(7) | Z,C,N,V | 1 |
| ROR | Rd | Rotate Right Through Carry | Rd(7) ← C, Rd(n) ← Rd(n+1), C ← Rd(0) | Z,C,N,V | 1 |
| ASR | Rd | Arithmetic Shift Right | Rd(n) ← Rd(n+1), n=0..6 | Z,C,N,V | 1 |
| SWAP | Rd | Swap Nibbles | Rd(3..0) ← Rd(7..4), Rd(7..4)←Rd(3..0) | None | 1 |
| BSET | s | Flag Set | SREG(s) ← 1 | SREG(s) | 1 |
| BCLR | s | Flag Clear | SREG(s) ← 0 | SREG(s) | 1 |
| BST | Rr, b | Bit Store from Register to T | T ← Rr(b) | T | 1 |
| BLD | Rd, b | Bit load from T to Register | Rd(b) ← T | None | 1 |
| SEC | | Set Carry | C ← 1 | C | 1 |
| CLC | | Clear Carry | C ← 0 | C | 1 |
| SEN | | Set Negative Flag | N ← 1 | N | 1 |
| CLN | | Clear Negative Flag | N ← 0 | N | 1 |
| SEZ | | Set Zero Flag | Z ← 1 | Z | 1 |
| CLZ | | Clear Zero Flag | Z ← 0 | Z | 1 |
| SEI | | Global Interrupt Enable | I ← 1 | I | 1 |
| CLI | | Global Interrupt Disable | I ← 0 | I | 1 |

| | | | | | |
|---|---|---|---|---|---|
| SES | | Set Signed Test Flag | S ← 1 | S | 1 |
| CLS | | Clear Signed Test Flag | S ← 0 | S | 1 |
| SEV | | Set Twos Complement Overflow | V ← 1 | V | 1 |
| CLV | | Clear Twos Complement Overflow | V ← 0 | V | 1 |
| SET | | Set T in SREG | T ← 1 | T | 1 |
| CLT | | Clear T in SREG | T ← 0 | T | 1 |
| SEH | | Set Half Carry Flag in SREG | H ← 1 | H | 1 |
| CLH | | Clear Half Carry Flag in SREG | H ← 0 | H | 1 |
| **Data Transfer Instructions** | | | | | |
| MOV | Rd, Rr | Move Between Registers | Rd ← Rr | None | 1 |
| MOVW | Rd, Rr | Copy Register Word | Rd+1:Rd ← Rr+1:Rr | None | 1 |
| LDI | Rd, K | Load Immediate | Rd ← K | None | 1 |
| LD | Rd, X | Load Indirect | Rd ← (X) | None | 2 |
| LD | Rd, X+ | Load Indirect and Post-Inc. | Rd ← (X), X ← X + 1 | None | 2 |
| LD | Rd, - X | Load Indirect and Pre-Dec. | X ← X - 1, Rd ← (X) | None | 2 |
| LD | Rd, Y | Load Indirect | Rd ← (Y) | None | 2 |
| LD | Rd, Y+ | Load Indirect and Post-Inc. | Rd ← (Y), Y ← Y + 1 | None | 2 |
| LD | Rd, - Y | Load Indirect and Pre-Dec. | Y ← Y - 1, Rd ← (Y) | None | 2 |
| LDD | Rd,Y+q | Load Indirect with Displacement | Rd ← (Y + q) | None | 2 |
| LD | Rd, Z | Load Indirect | Rd ← (Z) | None | 2 |

*(Continued)*

| Instr. | Operand | Name | Meaning | Flags | Cycles |
| --- | --- | --- | --- | --- | --- |
| LD | Rd, Z+ | Load Indirect and Post-Inc. | Rd ← (Z), Z ← Z+1 | None | 2 |
| LD | Rd, -Z | Load Indirect and Pre-Dec. | Z ← Z - 1, Rd ← (Z) | None | 2 |
| LDD | Rd, Z+q | Load Indirect with Displacement | Rd ← (Z + q) | None | 2 |
| LDS | Rd, k | Load Direct from SRAM | Rd ← (k) | None | 2 |
| ST | X, Rr | Store Indirect | (X) ← Rr | None | 2 |
| ST | X+, Rr | Store Indirect and Post-Inc. | (X) ← Rr, X ← X + 1 | None | 2 |
| ST | - X, Rr | Store Indirect and Pre-Dec. | X ← X - 1, (X) ← Rr | None | 2 |
| ST | Y, Rr | Store Indirect | (Y) ← Rr | None | 2 |
| ST | Y+, Rr | Store Indirect and Post-Inc. | (Y) ← Rr, Y ← Y + 1 | None | 2 |
| ST | - Y, Rr | Store Indirect and Pre-Dec. | Y ← Y - 1, (Y) ← Rr | None | 2 |
| STD | Y+q,Rr | Store Indirect with Displacement | (Y + q) ← Rr | None | 2 |
| ST | Z, Rr | Store Indirect | (Z) ← Rr | None | 2 |
| ST | Z+, Rr | Store Indirect and Post-Inc. | (Z) ← Rr, Z ← Z + 1 | None | 2 |
| ST | -Z, Rr | Store Indirect and Pre-Dec. | Z ← Z - 1, (Z) ← Rr | None | 2 |
| STD | Z+q,Rr | Store Indirect with Displacement | (Z + q) ← Rr | None | 2 |
| STS | k, Rr | Store Direct to SRAM | (k) ← Rr | None | 2 |
| LPM | | Load Program Memory | R0 ← (Z) | None | 3 |
| LPM | Rd, Z | Load Program Memory | Rd ← (Z) | None | 3 |
| LPM | Rd, Z+ | Load Program Memory and Post-Inc | Rd ← (Z), Z ← Z+1 | None | 3 |

| SPM | | Store Program Memory | (z) ← R1:R0 | None | - |
|---|---|---|---|---|---|
| IN | Rd, P | In Port | Rd ← P | None | 1 |
| OUT | P, Rr | Out Port | P ← Rr | None | 1 |
| PUSH | Rr | Push Register on Stack | STACK ← Rr | None | 2 |
| POP | Rd | Pop Register from Stack | Rd ← STACK | None | 2 |
| **MCU Control Instructions** | | | | | |
| NOP | | No Operation | | None | 1 |
| SLEEP | | Sleep | | None | 1 |
| WDR | | Watchdog Reset | | None | 1 |
| BREAK | | Break For On-chip Debug Only | | None | N/A |

Source: Atmel Corporation

## Flags

This section describes each flag bit and how and why they are used. The concept of a flag is the same for all microprocessors. Flags are used in loops and decisions (branches). A loop might be used to add up a series of numbers, to count, or to wait until a sensor is active. A branch can be used to implement an IF THEN ELSE structure. Before getting into those topics, we will first need to understand how each flag acts.

The SREG, status register, is an 8-bit flag register. The names of the flags are: I, T, H, S, V, N, Z, and C. Each bit in SREG represents a different flag or condition. Each flag is 1 bit, so the value is either a 1 or 0. At first, we will be concerned with the lower four flags: V, N, Z, and C. This next list states the names of the flags and what causes the flag to be set or equal to a 1.

V = Overflow. Set when the result exceeds the range of signed numbers.

N = Negative. Set when the result is negative.

Z = Zero. Set when the result is zero.

C = Carry. Set when the result exceeds the range of unsigned numbers.

The flags are continually being set according to the result of each instruction as the program runs. After an instruction produces a result, the flags are updated accordingly to that result. If, for example, an addition produces the result of zero, then the zero flag will be set. The next instruction after the add may look at the zero flag and then either branch or not branch according to the value of the zero flag. On the other hand, the programmer may choose to ignore the zero flag after the add depending on the intent of the programmer. If you take a look at the instruction set, you will notice how each instruction such as add or sub affects the flags. An instruction may or may not affect each flag, even though it does produce a result. One must refer to the instruction set. This next section of code shows an instruction, the result, and then the values of the flags.

```
clr    r16         ; r16 = 0. Thus z flag is set, n flag is not set
ldi    r16,1       ; ldi does not change any flags.
subi   r16,2       ; result = -1; Thus: N=1, Z=0, V=0, C = 1
ldi    r16,127     ; ldi does not change any flags
ldi    r17,3       ; ldi does not change any flags
add    r16,r17     ; result = 127+3 = 130 = -3; Thus: N=1, Z=0, V=1, C = 0
```

There is one caveat, however; one must check to see what effect, if any, each instruction has on the flags. For this detailed information, refer to the instruction set. For example, the instruction ldi will not affect the flags, while the instructions add or sub will affect V, N, Z, and C. The push and pop instructions will not change any flags either.

What are the flag values for these instructions after they run?

```
              V N Z C
clr    r17    0 0 1 0
ldi    r16,2  0 0 1 0
com    r17    0 1 0 1
neg    r16    0 1 0 1
dec    r17    0 1 0 1
subi   r17,1  0 1 0 0
```

What is the difference between dec r16 and subi r16,1? On the surface, it seems that they are the same, as they both subtract 1 from r16. But the subi affects the carry flag and the dec does not. So, on the occasions that r16 was initially 0 and using the dec instruction, one would not be able to check for a carry. If you want to subtract by 1, use the subi instruction. If you are counting loops down to zero, then use the dec instruction.

### Zero Flag (Z Flag)

The Z flag is set when the result is zero. Do not forget to check to see if the particular instruction affects the Z flag. If the instruction does not affect the flag, then that flag remains unchanged from the prior instruction.

```
ldi    r16,5    ; value of z flag is unchanged from prior instruction
subi   r16,2    ; z flag is 0
subi   r16,3    ; z flag is 1
```

### Negative Flag (N Flag)

The N flag is set when the result is less than zero. This flag interprets the result as a signed number even if you do not intend the result to be signed. If you are not working with signed numbers, then ignore this flag. A negative number has its most significant bit equal to a 1.

For example:

```
                                    N Z
ldi    r16,1      ; result = 1      ? ?
ldi    r17,2                        ? ?
add    r16,r17    ; result = 3      0 0
subi   r16,3      ; result = 0      0 1
subi   r16,1      ; result = ff = -1  1 0
```

After lines 1 and 2, how are the flags set? The instruction ldi will not affect any flags, and we do not know what they were prior to these lines. After line 3, how are the flags set? The add instruction will affect N, Z, V, and C. The add produced a result of 3. The N flag is not set (0) because the result was not negative. The Z flag is not set (0) because the result was not zero. Also note that flags V and C are not set because one does not exceed the range of signed or unsigned numbers, respectfully.

After subi r16,3, how are the flags set? The add will affect N, Z, V, and C. The result is $3 - 3 = 0$. The N flag is not set (0) because the result was not negative. The Z flag is set (1) because the result was zero.

After subi r16,1, how are the flags set? The result is $0 - 1 = -1$. The N flag is set (1) because the result was negative. The Z flag is not set (0) because the result was not zero.

### Carry Flag (C Flag)

The C flag is set when the result of an instruction exceeds the range for unsigned integers. The range of unsigned numbers using 8 bits is 0 to 255. If the computation crosses the

[0,255] boundary, then the C flag is set. You can ignore this flag if using signed numbers. For example, $253 + 5 = 258 = 256 + 2$, and thus the result is 2 and a carry. The value of the carry is 256. When doing addition or subtraction and when there is a possibility that the result may go beyond 255, then check for a carry. For example:

```
                      ;result    N Z V C
    ldi   r17,4
    ldi   r18,2
    ldi   r19,1
    ldi   r16,250   ; 250 = -6
    add   r16,r17   ; 254 = -2      1 0 0 0
    add   r16,r18   ;   0 =  0      0 1 0 1
    add   r16,r19   ;   1 =  1      0 0 0 0
    subi  r16,2     ; 255 = -1      1 0 0 1
```

### Overflow Flag (V Flag)

The V flag is set when a result exceeds the range for signed integers. Using 8 bits, the range of numbers is $-128$ to $+127$. In other words, if the computation crosses the $[127,-128]$ boundary, the V flag is set. The decimal range $[-128,127]$ is the same as the hex range $[80,7F]$. The hex range is good to use sometimes because you can see the most significant bit (MSB) more easily. Negative numbers have a MSB equal to 1. You can ignore this flag if you are not using signed numbers.

If register r16 = 255 and we add 1 to get 0, what flags are set? The Z flag is set. The N flag is not set. The carry flag is set because the range for unsigned numbers has been exceeded. The V flag is not set. The reason for this is that 255 (ff hex) as a signed number is $-1$ and when 1 is added you get 0. In other words, $-1 + 1 = 0$ and this does not exceed or go outside the range of $-128$ to $+127$.

If register r16 = 125 and we add 2 to get 127 (7f hex) then the N, Z, V, and C flags are not set. But if we add 1 more we get 128 (80 hex). In this case, the N and V flags are set while Z and C are not set. The V flag is set and indicates that the result has exceeded the range of signed numbers. If you are not using signed numbers, then ignore the V flag and interpret the result as $127 + 1 = +128$. Notice how the 7f went to 80, which is natural, and that the MSB is set.

Another example is if r16 = $-100$ and we subtract 30, then the result is $-130$, which does not exist. The number $-130$ is 2 beyond $-128$, which is 126. Note that 1 to the left of $-128$ is $+127$, so 2 to the left of $-128$ is $+126$. After the subtraction, the V flag will be set.

The microprocessor does not know whether the numbers are signed or unsigned. The microprocessor just sets the flags according to the instruction set rules. It is the responsibility of the programmer to know which flags to interpret. These next instructions directly affect the flags.

| Flag Instruction | Meaning |
| --- | --- |
| clc | clear the carry (make 0) |
| sec | set carry (make 1) |
| clv | clear v |
| sev | set v |
| cln | clear n |
| sen | set n |
| clt | clear t |
| set | set t |
| cli | global interrupt disable |
| sei | global interrupt enable |

You can run a program line-by-line using the simulator. This is called a trace. After executing a line, the computer responds with the results and flag values. The 8-bit SREG register represents the flags. The lower 4 bits are the VNZC. So if the SREG = 8C, then VNZC = $1100_2$.

The upper four flags (I, T, H, S) of the SREG are used less often. They are as follows:

**I**: *Global Interrupt Enable*. This bit must be set to 1 for the interrupts to be enabled.

**T**: *Bit Copy Storage*. This can be a 1-bit storage by using instructions BLD and BST.

**H**: *Half-carry*. This is set when there is a carry out of, or borrow into, bit 3 of a result.

**S**: *Sign Bit*. The S-bit is the negative flag (N) exclusively ORed with the overflow flag (V).

## Looping

Looping is a common programming task. It is used to do such things as delaying, moving a list of numbers, adding up a list of numbers, or repeating a task. Often you know how many times to loop. In this case, simply load a register with the number of times to loop, and then in a loop decrement the register. When the register equals zero the looping ends.

### Example 1: Loop 5 Times

Before the loop, load r16 with 5. In the loop, decrement r16. At the bottom of the loop, use BRNE. BRNE means to branch if not equal, or in other words branch to the label (top) if the previous result (r16) is not equal to zero. The first time through, r16 decrements to 4, so it is not equal to zero and thus it will branch to the label.

Labels are placed in the very first column and end with a colon. Here the computer will execute the dec 5 times. When r16 becomes 0, the brne will not branch and the computer will go on and execute the nop instruction.

```
        ldi     r16,5
top:    dec     r16
        brne    top
        nop
```

What flag was used here with the brne instruction? The zero flag. How many times would it loop if the dec r16 was changed to inc r16? Would it go on forever? The answer is that r16 would count up to 255 and then count 1 more to reach 0 and at that point the loop would end. This means the instruction inc r16 occurs 251 times.

```
        ldi     r16,5
top:    inc     r16
        brne    top
```

### Example 2a: Loop 1,000 Times

This is a bit harder. Only a 16-bit register can hold a number that large. An 8-bit register can go up to 255, while a 16-bit register can go up to 65535. The answer uses a loop within a loop. This allows the body of the loop to run many times, although the code is more complex, and complex code is harder to debug and is more costly. Simple, straightforward code is easier to maintain. In this example, the inner loop runs 100 times and the outer loop runs the inner loop 10 times, for a total of $10 \times 100$, or 1000 times.

```
        ldi     r16,10
topb:   ldi     r17,100
topa:   dec     r17
        brne    topa
        dec     r16
        brne    topb
```

## Example 2b: Loop 1,000 Times

Another solution uses the x register, which is xh and xl concatenated to produce a 16-bit register. A 16-bit register can count up to 65535. The difficulty here is in loading x. One has to load xl and xh with the value of 1000. Since xh is the high byte of xh:xl, its lowest bit has a value of 256. So how many 256's go into 1000? That is simply 1000/256. The remainder, 1000%256, goes into xl. Also, note that instead of using a dec instruction a sbiw is used. A dec can decrement xl or xh individually but not xh:xl together as needed.

```
        ldi     xh,1000/256
        ldi     xl,1000%256
top:
;    some task goes here
        sbiw    x,1
        brne    top
```

## Example 3: Read port b until portb Is 0x0f

Here we want to loop until portb is 0f (hex); thus, we do not know how many times it will loop. In the loop, read the port and compare it to 0x0f. Use a brne (branch if not equal) to 0x0f. The instruction cpi compares a register with a number but does not change the register. Remember that portb is the name of the port when doing output and pinb is the name of the port when doing input.

```
again:  in    r16,pinb
        cpi   r16,0x0f
        brne  again
```

## Example 4: Read port b until the First 2 Bits are Set

This is similar to the prior problem with the exception that we do not care about the upper 6 bits. In addition, we do not know how many times to loop. In the loop, read the port, mask out the upper 6 bits to zero, and then compare it to 0b00000011. Since we do not care about the upper 6 bits and we do not know what they will be, we must mask them out to known values, like 1 or 0.

```
again:  in    r16,pinb          ; read port
        andi  r16,0b00000011    ; set upper bits to 0
        cpi   r16,0b00000011
        brne  again
```

By the way, the andi does affect the zero flag, which means that the cpi is redundant and is not needed for the brne.

### Example 5: Compute the Summation: 10+9+8+ ... +1

When finished, r16 will contain the answer. The register r17 will be the counter that goes from 10 down to 1. What is the summation? Will it fit into r16? The summation is 11 × 5, or 55.

```
        ldi     r16,0       ; summation
        ldi     r17,10      ; 10 values
more:   add     r16,r17     ; add number into r16
        dec     r17
        brne    more
```

## Jumping

A jump sends the program counter to the location of the label. The operand after the jump is the name of the label and also the address of where to go to. The instruction rjmp is a relative jump and can be used on the ATtiny13, ATtiny85, ATmega328, and ATmega32. The instruction jmp is a direct jump and can be used on the ATmega328, and ATmega32. A jmp can be located further away from the label than an rjmp can be located.

```
        rjmp    xyz
                ...
        jmp     xyz     ; ATmega328 or ATmega32
                ...
xyz:            ...
```

## Branching

Normally a program executes from top to bottom, one instruction at a time. Branching and jumping allow the instruction sequence to change or branch to another location. The location to branch to is identified by a label. Labels end with a colon. A branch may be conditional or unconditional. Unconditional branching is like a GOTO statement. Conditional branching is used to make a decision and thus possibly skip over lines of code. This is similar to an IF THEN ELSE statement. If the branch occurs, the next instruction to be executed is on the same line as the label.

A label is a destination and typically starts in the first column of the program. It can be preceded by a white space. The label is an address. It is the address of where to branch to. The label must be a unique name not used as any other label. It is helpful to use a meaningful name. While a label is an address and programmers can use addresses instead of labels, labels are much easier and clearer.

Branches are relative. The displacement (distance) of the branch is measured from the instruction after the branch to the label. Backward branches have a negative distance, while a forward branch has a positive distance. The label is the actual address of the destination. For any branch instruction, the compiler uses the label to compute the distance to the destination. This signed displacement can be negative. A one-byte signed number has the range of −128 to +127.

### Unconditional Branches

An unconditional branch will always branch. This instruction is followed by the name of the label to which to branch. The instruction rjmp is an unconditional relative jump.

### Conditional Branches

The conditional branch instruction will branch when a certain condition is met. Remember that the flags are set after each instruction. The conditional branch will either branch or not branch according to the flag values as set by the prior instruction. For example, a conditional branch instruction like breq checks to see if the previous instruction has a result equal to zero (zero flag set), and if it is, then it will branch to the specified label.

What does the following code do if r16 is initialized to 30, and what is the value of r17?

```
        ldi    r16,30
        clr    r17
top:    subi   r16,5
        breq   done
        inc    r17         ;count the number of loops
        rjmp   top
done:
```

Solution: 5 will be subtracted from r16 during each loop. When the subtraction reaches zero, the branch to done occurs. The loop counter, r17, will be 6. What would happen if r16 was initialized to 31? The result is a very long loop. Remember that branch works off of the flags, and the flags are set by the prior instruction result.

Following is a summary of the branch instructions. Use the signed branches when the values tested are signed numbers, and likewise use the unsigned branches when the values tested are unsigned. In general, use the unsigned branches when possible as they are easier to work with. Here is a table of simple conditional branches.

| Simple Conditional Branch Instruction | Meaning |
| --- | --- |
| breq | branch if equal |
| brne | branch if not equal |
| brcc | branch if carry clear |
| brcs | branch if carry set |
| brvc | branch if overflow clear |
| brvs | branch if overflow set |
| brpl | branch if plus, n clear |
| brmi | branch if minus or negative, n set |

Signed branches are used when working with signed numbers. For example, 0xff, as signed, is less than 0x7f because 0xff as signed is $-1$ and 0x7f as signed is 127. Very often, a compare precedes the branch instruction.

| Signed Branch Instruction | Meaning |
| --- | --- |
| brge | branch if >= |
| brlt | branch if < |

Unsigned branches are used when working with unsigned numbers. For example, 0xff is greater than 0x7f because 0xff as unsigned is 255, while 0x7f as unsigned is 127. Unsigned branches are easier to use and understand than signed branches.

| Unsigned Branch Instruction | Meaning |
| --- | --- |
| brsh | branch if >= |
| brlo | branch if < |

Most branches tend to be unsigned rather than signed. Unsigned measurements include counting items, distance, speed, money, weight, pressure, etc. Signed values include displacement, temperature, velocity, etc. Following are a few examples.

What will this code do? It will loop forever.

```
tip:    ldi    r16,0
        breq   tip
```

This next bit of code will read port b while the port is zero. In other words, it will read the port while all the bits are zero. A compare is not needed.

```
top:    in     r16,pinb
        breq   top
```

This code will branch to skip if r16 is not equal to 5.

```
        cpi    r16,5
        brne   skip
               ...
skip:
```

Branching is usually conditional (dependent on the previous result). If the branch is true, then the next instruction will be the one specified by the label. Remember that at the time of the branch the PC is pointing to the instruction on the next line after the branch. If the branch is false, then the branch label and offset do not matter and the MPU will execute the instruction on the next line after the branch.

Which lines branch?

```
        ldi    r16,0x81
        cpi    r16,0x02
        brge   me1          ; does not branch
        brlt   you1         ; does branch
        brsh   you2         ; would have branched if gotten here
me1:
you1:
you2:
```

Here are more examples of branching. Try to predict how many times each loop loops. Notice that the branch works according to the flag values that are set in the prior instruction. The name of each label must be unique.

```
        ldi    r16,2        ; 2 loops
belie:  dec    r16
        brne   belie
```

```
            ldi    r16,255    ; 1 loop
bucolic:    inc    r16        ; 255 + 1 is 0
            brne   bucolic

            ldi    r16,2      ; 3 loops
replete:    inc    r16
            cpi    r16,5
            brne   replete

            ldi    r16,0      ; adds up the numbers 3 to 13.
            ldi    r17,3
caveat:     add    r16,r17
            inc    r17
            cpi    r17,14     ; We are done when r17 = 14
            brne   caveat

callow:     in     r16,PINA   ; read port A until bit 0 = 1
            andi   r16,0b01
            breq   callow
```

Try these problems on branching and looping.

1. Load r16 with Port B. If r16 is greater than 0x59 (unsigned), then make r17 equal to 0xff; otherwise, clear r17.

```
            in     r16,pinb
            cpi    r16,0x5a
            brsh   skip        ; branch if >= unsigned
            ldi    r17,0xff
            rjmp   done
skip:       ldi    r17,0x00
done:
```

2. Load r16 with Port B. If r16 is negative, then make r17 equal to 0xff; otherwise, clear r17. The solution is the same as the prior code but with this branch.

```
            brpl   skip
```

3. Load r16 with any number you wish. If r16 has bits 0 and 2 set, then make r17 equal to 0xff; otherwise, clear r17.

```
            in     r16,pinb
            andi   r16,0b101
            brne   skip
            ldi    r17,0xff
            rjmp   done
skip:       ldi    r17,0x00
done:
```

4. Write the code. If r16 >= 0x12, then make r16 = r16 − 2. Treat r16 as unsigned. You need to use an unsigned branch.

```
        cpi     r16,0x12
        brlo    skip
        subi    r16,0x02
skip:
```

5. Write the code that will read port c until bit 5 becomes a 1 and bit 6 is a 0.

```
top:    in      r16,pinc
        andi    r16,0b01100000
        cpi     r16,0b00100000
        brne    top
```

## Direct Addressing

You can store data in the SRAM memory. Refer to the memory map for each AVR device to see where free memory is. For the '328, free volatile SRAM memory is from 0x0100 to 0x08ff. But remember that the stack memory starts at 0x08ff and grows toward lower memory. Don't use this stack area. The problem is that you can never be sure how much memory the stack uses, so if you need to use the SRAM memory, start at the other end of memory (0x0100). Remember that this memory is volatile, meaning that when the power is removed the data is lost. The instructions to access the SRAM memory are shown below.

| Direct Addressing Instruction | Meaning |
| --- | --- |
| sts address, register | Stores register value in memory at that address |
| lds register, address | Loads register value from memory at that address |

In this example, notice the use of the equate. It can make referencing memory locations easier.

```
.equ    table = 0x0100
.org    0x0000
    ldi     r16,0x16
    sts     table,r16        ; store to sram
    lds     r17,table        ; load from sram
    nop
```

## Indirect Addressing

Indirect addressing is a powerful yet more complicated way to access memory. Indirect addressing allows the programmer to easily access or manipulate a list of memory locations within a loop. Accessing a list of data would be much harder without indirect addressing. Indirect addressing uses an index register, which is the pointer to the data in memory. There are three index registers—x, y, and z—and each is 16 bits. For example, register x is the concatenation of xh:xl, and this also means x is register r27:r26 put together. Likewise, register y is r29:r28, and z is r31:r30.

Typically, one loads the index register with an address so it points to the beginning of a list of numbers and then uses a loop to go through the list of numbers that does something to or with the numbers. Within the loop, the index register needs to be incremented or decremented. This example puts the number 13 in each memory location from 0x0111 to 0x1f2 inclusive:

```
        Load x with starting address of data, 0x0111
        Load r16 with 13
top:    Store r16 at location pointed to by x, then increment x.
        Does x point to one past the end of the list (0x01f3)? If not, go to top.
done:
```

Of course, you need to convert the preceding ideas into actual instructions. The instructions to access data (in SRAM memory) using indexed addressing are ld and st. The instruction ld means load from memory, while st means store to memory. Converting the example to actual code produces the following:

```
      ldi    r16,13
      ldi    xl,0x11      ; initialize pointer
      ldi    xh,0x01
top:
      st     x+,r16       ; store value then move pointer
      cpi    xl,0xf3      ; are we done?
      brne   top
```

You can automatically increment or decrement the index register by 1. A + or − before the register x, y, or z is called *pre*-increment or *pre*-decrement and is done before the instruction executes. A + or − after the register x, y, or z is called *post*-increment or *post*-decrement and is done after the instruction executes. The programmer needs to pay attention to the value used in the cpi with respect to how and when the index register moves. Here are the various index loads and stores.

| Indirect Addressing Instruction | Meaning |
|---|---|
| LD    Rd,Z | Load Indirect |
| LD    Rd,Z+ | Load Indirect and Post-Increment |
| LD    Rd,-Z | Load Indirect and Pre-Decrement |
| LDD   Rd,Z+q | Load Indirect with Displacement |
| ST    Z,Rr | Store Indirect |
| ST    Z+,Rr | Store Indirect and Post-Increment |
| ST    -Z,Rr | Store Indirect and Pre-Decrement |
| STD   Z+q,Rr | Store Indirect with Displacement |

Instead of accessing the SRAM, the programmer can access the Flash memory. A nice feature about the Flash memory is that the values are retained even after the power is removed. To access the Flash memory where programs and non-volatile memory is stored, use the instructions lpm and spm as shown below.

| Indirect Addressing Instruction | Meaning |
|---|---|
| LPM | Load Program Memory |
| LPM   Rd,Z | Load Program Memory |
| LPM   Rd,Z+ | Load Program Memory |
| SPM | Store Program Memory |

Referring back to the prior example, what if the data locations range from 0x0111 to 0x02f2? Note that xh starts at 0x01 and ends with the value 0x02. What needs to change is the compare because it is no longer sufficient to end the loop looking only at xl. The solution is to check both xl and xh for final values. This code stores a 13 in memory locations 0x0111 to 0x02f2.

```
ldi    r16,13
ldi    xl,0x11    ; starting memory location
ldi    xh,0x01
```

```
top:
    st      x+,r16      ; store a 13, then increment x.
    cpi     xl,0xf3     ; is xl done?
    brne    top
    cpi     xh,0x02     ; is xh done?
    brne    top
```

This next example uses indexed addressing to add a column of 1024 bytes starting at memory location 0x0100. It puts the result into the two registers r17:r16.

```
    ldi     xl,0x00
    ldi     xh,0x04     ; x = # of numbers to work on.

    ldi     yl,0x00
    ldi     yh,0x01     ; y = starting memory location of data.

    clr     r16         ; result low   byte
    clr     r17         ; result high byte
    clr     r19         ; always zero
top:
    ld      r18,y+      ; get number
    add     r16,r18     ; add to low byte
    adc     r17,r19     ; add any carry to high byte. r19 is zero.

    sbiw    x,1         ; go through list of 1024 values
    brne    top
```

The following example illustrates how you can search through a list of numbers looking for a certain value. In this case, the program is looking for the lowest value. The x index register is used to point to the data, while y is used to keep track of how many numbers were looked at.

```
; ----------------------------
; ATmega328. Finds the lowest value in the 600 memory locations
;     starting at 0x0150. Answer is stored in r16.
.org    0x0000
    ldi     xl,0x50         ; start of data
    ldi     xh,0x01
    ldi     yl,600%256      ; # of data values
    ldi     yh,600/256
    ld      r16,x+          ; get first data value
    sbiw    y,1             ; decrement count
top:
    ld      r17,x+          ; get new value to compare
    cp      r16,r17         ; compare r16 to new value
    brsh    skip            ; branch if >=
    mov     r16,r17         ; remember this new lowest
```

```
skip:
    sbiw   y,1                ; decrement count
    brne   top
done:
    rjmp   done
; - - - - - - - - - - - - - - - - - - - - - - - - - - - - -
```

The following program moves data from one block of memory to another. The source is unchanged. Notice that the use of equates makes the program easier to understand and change. It is interesting that xl is found from the remainder of 0x0200 % 256. The % symbol yields the modulus (remainder). And xh is 0x0200 / 256 which leaves out the remainder. The post increment x+ increases x after it is used in the instruction.

```
; ATmega328. Move the data found in memory location [0x0200,0x03ff]
;    to memory location [0x0400,0x05ff].
.equ    source_start         = 0x0200
.equ    source_end           = 0x03ff
.equ    destination_start    = 0x0400

.org    0x0000
    ldi   xl,source_start % 0x0100          ; start of data source
    ldi   xh,source_start / 0x0100
    ldi   yl,destination_start % 0x0100    ; start of data destination
    ldi   yh,destination_start / 0x0100
more:
    ld    r16,x+                    ; pick up data, then move x.
    st    y+,r16                    ; put down data, then move y.
    cpi   xl,(source_end+1) % 0x0100        ; are we done?
    brne  more
    cpi   xh,(source_end+1) / 0x0100
    brne  more
done:
    rjmp  done
; - - - - - - - - - - - - - - - - - - - - - - - - - - - - -
```

## Stack Memory

The stack is an area of memory used for storing temporary data. The stack is also needed for subroutines. Subroutines use the stack automatically without the programmer explicitly using stack instructions. You generally do not need to initialize the stack pointer.

SRAM memory can be accessed in random order by using the instructions lds or sts, but this is not so with the stack. The stack instructions access the SRAM memory in sequential order. The stack is similar to a stack of dishes on a table. When you save the first number to the stack, it is analogous to placing a plate on the table. Save another number (or plate) and this goes on top of the first one. Save a third, and now we have 3 numbers (or plates) stacked up. The numbers in the stack are stored consecutively in memory and the list of numbers grows closer toward lower memory. So numbers can be pushed onto the stack and at any time they can be pulled off, but to get to the first number (or bottom plate), one first has to pull off the topmost numbers (or plates), one at a time. This type of stack is called LIFO, or Last-In First-Out. Immediate access is limited to only the topmost item. You can either push a number onto the stack or take the top number off.

To keep track of the last byte pushed onto the stack, there is a pointer called the stack pointer (SP). Normally you should not be concerned with the value of the SP, but do remember that on the AVR the SP points to the next available unused byte (see the following figure). The stack on the AVR is initialized to the last available byte in SRAM. On the ATmega328, this is memory location 0x08ff. Thus, you need not initialize the SP, but if you need to you can. For the AVR, as you push data onto the stack, the stack grows toward lower memory addresses. The AVR's SP is 16 bits. The register SPH is the upper 8 bits, while the register SPL is the lower 8 bits of SP.

The syntax for pushing register r16 onto the stack is: push r16

The syntax for pulling register r16 from the stack is: pop r16

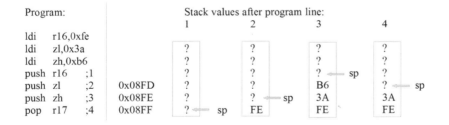

This example demonstrates how to swap r16 and r17 using the stack.

```
push   r16
push   r17
pop    r16
pop    r17
```

What is the limit to our example with the plates? You can place plates onto the stack until you run into the ceiling. A similar thing happens with the stack on the AVR. Here, the limitation is available memory space. Generally, there is no warning whatsoever when the computer stack overflows, and it will most likely crash at that point. On the AVR, data is stored at the start of the SRAM, while the stack is at the end of the SRAM. These two areas may grow and could run into each other, at which time the AVR will crash. This condition is called an overflow.

Another potential problem is an underflow of data. For every push, there should be one pop. An underflow occurs when one attempts to pop off more numbers from the stack than were on the stack originally. Additionally, if you forget to pull a number off the stack, then the next time you pull from the stack it will be the wrong number. This last issue can cause problems when returning from a subroutine. Try the following problems for practice.

```
;What are the values of the registers after each line:
    ldi     r16,33          ;33
    ldi     r17,44          ;44
    subi    r16,0b11        ;30
    push    r16             ;r16 is unchanged
    push    r17             ;r17 is unchanged

    ldi     r16,33          ;33
    ldi     r17,44          ;44
    push    r16             ;no change
    dec     r16             ;32
    pop     r17             ;33

    ldi     r16,16          ;16
    push    r16             ;no change
    push    r16             ;no change
    pop     r17             ;16
    pop     r18             ;16
    add     r17,r18         ;32
```

Stacks can be thought of as either linear or circular, depending on the implementation and application. A linear stack has one SP, and it points to the end of the stack, which grows or contracts. The data last in is the data first out (LIFO). Applications include storing data, making registers local to a subroutine, and passing data to a subroutine.

Linear Stack - LIFO

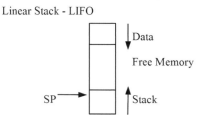

A circular stack is organized as First-In First-Out (FIFO). An example of this would be a print queue. The first print job submitted should be printed before subsequent print requests. To handle this, two pointers are needed: one SP for the beginning of the list, and another SP for the end of the list. If the list gets too long, it can overflow the allotted memory. At this point, the two stack pointers collide. In this picture, the tail SP increments when a print job is finished, while the head SP increments when a print job is submitted.

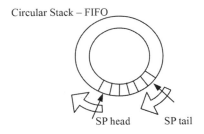

Circular Stack – FIFO

SP head        SP tail

## Subroutines and Delays

Subroutines are sections of code that typically do one task. They can be located anywhere in the program but are generally placed at the bottom of a program or in a separate file. Once the subroutine is completed and tested, it should be kept safe from changes—preferably in a separate file. Furthermore, those finished subroutines do not need to be near or next to code that is being written and debugged for fear of being corrupted.

Typical tasks for a subroutine include delay routines, mathematical equations, sorting, checking sensors, or any other task that may be used frequently. Subroutines will make your program easier to write, debug, and understand by breaking up a complex problem into smaller tasks. Also, you can build a collection of routines that you will be able to use in future programs. The syntax of a subroutine is straightforward. In the main program, issue a jump to the subroutine by using the instruction rcall followed by the name of the subroutine. This will run the subroutine by that name. When the subroutine is to finish, issue the return instruction ret. Here is an example of a routine that performs a delay. Delay calculations are shown later, but essentially the delay time is equal to the number of loops times the time per loop.

```
.org     0x0000
         ...
         rcall      delay
         ...
```

```
; ----------------------
delay:  push       r16
        ldi        r16,255
delay1: nop
        dec        r16
        brne       delay1
        pop        r16
        ret
; ----------------------
```

In the prior code, you will notice the subroutine follows the main program. When the instruction `rcall` occurs, the AVR pushes onto the stack the return address (the low byte is pushed first, followed by the high byte). The return address is the address of the instruction immediately after the `rcall` instruction. This return address is the address to return to when the subroutine is finished. At the end of the subroutine, the `ret` instruction executes and the AVR pulls off the stack the return address, goes to that return address, and continues the program. From this statement, you should then realize that the stack must not be corrupted in the subroutine by having an unequal number of pushes and pulls. Each subroutine must have a unique name or label.

It is a good idea to push and pull each register used in a subroutine. This is good practice because then it does not burden the main program with the possibility that the subroutine will destroy the register values used by the main program. For every push, there should be a pop. Consider why it is a mistake when a programmer pushes r16 at the beginning of a subroutine but does not pop it off at the end of the subroutine. The answer is that the `ret` instruction will not get the correct return address from the stack, and consequentially the program will crash.

Can a subroutine call another subroutine? Yes, it can. The only limitation is the available stack size. In fact, a subroutine may call itself, which is called *recursion*. Recursion can be a very powerful technique for solving some problems. A simple example is in computing a factorial. The factorial of $5 = 5! = 5 \times 4!$, where $4! = 4 \times 3!$, where $3! = 3 \times 2!$, where $2! = 2 \times 1!$, and $1! = 1$.

It is recommended to keep your programming simple, straightforward, and uncomplicated. When making a label within a subroutine, a good idea is to use the same name as the subroutine and then append a number. This way all of the label names will be unique. The name "top" is a common label, but remember that there cannot be two labels with the same name. A few names may not be used as a label. For example, it is not possible to use the name of an instruction as a label name.

**Example 1.** This subroutine shows how to add up the registers r1, r2, r3, r4, r5 and put the answer in r0. Push and pop are not used since none of the registers are being altered except r0, which must be altered.

```
addup:
    clr    r0
    add    r0,r1
    add    r0,r2
    add    r0,r3
    add    r0,r4
    add    r0,r5
    ret
```

**Example 2.** For a longer delay, you could add more nop instructions in the loop, but it is probably better to use a 16-bit register like x, y, or z for the loop counter. The 16-bit register will have to be loaded 8 bits at a time, but the subtraction, sbiw, works on the entire 16 bits. Notice the pushes and pops, along with the label names.

```
delay2:   push   zl            ; For a longer delay
          push   zh
          ldi    zl,0xff
          ldi    zh,0xff
delay21: nop
          sbiw   z,1
          brne   delay21
          pop    zh
          pop    zl
          ret
```

You can determine how long a program will take to execute as each instruction takes a set prescribed time to execute. The time for an instruction is given in cycles, and the time for a cycle is set by the clock frequency. The frequency for the ATmega328 and ATmega32 is generally set at 1 MHz, while the ATtiny13 is set at 1.2 MHz. At 1 MHz, a one-cycle instruction will take 1 microsecond, but keep in mind that the internal clock can vary slightly.

**Example 3.** Write a subroutine that will take 1 msec if the frequency is 1 MHz. The solution starts with computing the number of cycles needed. 1 msec = (.001 sec) × 1,000,000 cycles/sec = 1000 cycles. Use a loop to get 1000 cycles. Most instructions take 1 cycle. The code is shown below. How many loops is 1000 cycles? First, make a loop and get the cycle count per loop. One way to do that is to look at the instructions listing file from AVR Studio or the instruction set to determine the number of cycles for each instruction. This loop is 10 cycles long. 10 is a very nice number; this is done on purpose! The brne takes 2 cycles when it branches and 1 cycle when it does not branch. 1000 cycles × (1 loop/10 cycles) = 100 loops.

```
delay_1ms:
        push    r16         ; [2]
        ldi     r16,100     ; [1]
delay_1a:
        nop                 ; [1]
        nop                 ; [1]
        nop                 ; [1]
        nop                 ; [1]
        nop                 ; [1]
        nop                 ; [1]
        nop                 ; [1]
        dec     r16         ; [1]
        brne    delay_1a    ; [2]
        pop     r16         ; [2]
        ret                 ; [4]
```

**Example 4.** The next bit of code shows how you can make a LIFO stack for storing values in register r16. The index register z was chosen as a stack pointer. First, the main program initializes the stack pointer, z. The stack starts at 0x06ff and grows toward low memory. The push_r16 subroutine uses a pre-decrement, and for this reason, the stack is initialized to 0x700.

```
;;Main program. Set up stack pointer z
    ldi     zl,0x00
    ldi     zh,0x07

    ldi     r16,0x33
    rcall   push_r16
    rcall   push_r16

    ...

; — — — — — — — — — — — — — — — — — —
;; This will take r16 and push it onto a LIFO
;;   stack pointed to by z
push_r16:
    st      -z,r16
    ret
; — — — — — — — — — — — — — — — — — —
; — — — — — — — — — — — — — — — — — —
```

```
;; This will take r16 and pop it from a LIFO
;;    stack pointed to by z
pop_r16:
    ld      r16,z+
    ret
; ---------------------
```

The prior example does not check the bounds of the pointer, which can be an issue if one runs out of memory space. Also, the pointer z is a global variable, which means that the code is non-reentrant.

**Example 5**. How would the prior example change if you wanted to build a circular FIFO stack? For a picture of what a circular stack looks like, refer to the prior section on the stack. To begin with, you would need two pointers: a pointer_head and a pointer_tail. When pushing or popping you should check to see that:

- The memory space available is not exceeded (head runs into the tail).
- Not too many items are pulled off of the stack (tail runs into the head).
- When a pointer reaches the end of the linear memory range, it needs to wrap around to the other end of memory. This makes a linear region look like a circular region.

**Example 6**. How does one make a delay longer than the prior examples? One way to do it is to make a loop within a loop. However, loops within loops are harder to calculate and adjust to get an exact delay time. Instead of a loop within a loop, consider a subroutine of, say, 1000 loops, and in the loop call another subroutine of a given short delay such as 1/100 sec, thus producing a delay of about 1000/100 sec or 10 sec. Perform the delay calculation for this subroutine.

```
; A delay subroutine using a loop within a loop.
del:
    push    r20
    push    r21
    ldi     r20,0xff
del1:
    ldi     r21,0xff    ;outer loop
del2:
    dec     r21         ;inner loop
    brne    del2        ;inner loop
    dec     r20         ;outer loop
    brne    del1        ;outer loop
    pop     r21
    pop     r20
    ret
```

**Example 7.** When writing larger programs, make your subroutines, test and debug them, and then put them away out of sight and safe from accidental changes. Do this by saving them in a separate file for use by this program and future programs. For the next section of code, the delays needed by the main program will be incorporated by using .include as shown. The include is at the bottom of the main program. The file to be included is in the same directory as the main asm program.

```
.org    0x0000
        ...
        rcall       delay_1s
        ...
.include <delays_1mhz.asm>    ;includes the code in this file
```

These delay subroutines are designed to run at 1 MHz.

```
; — — — — — — — — — — — — — — — — — —
; delays_1mhz.asm - Delays at 1 MHz
;delay_1s            ;1 second delay
;delay_100ms         ;100 millisecond delay
;delay_5ms           ;5 millisecond delay
;delay_1ms           ;1 millisecond delay
; — — — — — — — — — — — — — — — — — —
;1s delay
delay_1s:
        push        r16
        ldi         r16,200
delay_1s1:
        rcall       delay_5ms
        dec         r16
        brne        delay_1s1
        pop         r16
        ret
; — — — — — — — — — — — — — — — — — —
; — — — — — — — — — — — — — — — — — —
;1/10s delay
delay_100ms:
        push        r16
        ldi         r16,100
delay_100ms1:
        rcall       delay_1ms
        dec         r16
        brne        delay_100ms1
        pop         r16
        ret
```

```
; ————————————————————— —
; —————————————————————
delay_5ms:
    rcall    delay_1ms
    rcall    delay_1ms
    rcall    delay_1ms
    rcall    delay_1ms
    rcall    delay_1ms
    ret
; —————————————————————
; —————————————————————
;10us/loop
delay_1ms:
    push     r16
    ldi      r16,99 ;accounts for overhead of 12 cycles.
delay_1ms1:
    nop
    nop
    nop
    nop
    nop
    nop
    nop
    dec      r16
    brne     delay_1ms1
    pop      r16
    ret
; —————————————————————
```

Consider this example code. How long does this code take to execute? What percent of the time is the code within the loop? What value of zh:zl would give the longest delay?

```
    ldi    zh,4    [1]
    ldi    zl,0    [1]
top: nop          [1]
    sbiw   z,1     [2]
    brne   top     [2]
```

There are 4 × 256 loops and 5 cycles per loop. This yields 1024 cycles in the loop, plus 2 cycles outside the loop. The last brne is 1 cycle, not 2. The total is 1024 + 2 − 1 = 1025 cycles. The time in the loop is 1023 out of 1025 cycles. The longest delay would be when zh:zl is 0xffff. Actually, the value would be 0x0000 because there is a subtract before the branch.

What value of zh:zl would yield a delay of 1 msec? At 1 MHz, or 1,000,000 cycles per second, 1000 cycles would take 1 msec. So the value of zh would be 3, and zl would be 251.

Will this subroutine return properly to the calling program?

```
sub1:   push    r0
        push    r1
        ...
        pop     r16
        ret
```

No, it will not, due to the fact that 2 bytes are pushed onto the stack but only 1 is pulled off before the ret instruction.

Write the code to blink on and off all LEDs connected to port b every 1/2 second.

```
.org  0x0000
        ldi     r17,0x00
        ldi     r16,0xff
        out     ddrb,r16
top:    out     portb,r17
        rcall   delay_half_sec
        out     portb,r16
        rcall   delay_half_sec
        rjmp    top
```

Write the code to make port b an output port. Then send to port b the numbers 0 to 59 (decimal) and repeat. Each number is displayed for 1 sec.

```
.org  0x0000
        ldi     r16,0xff
        out     ddrb,r16
restart:
        ldi     r17,0x00
top:    out     portb,r17
        rcall   delay_sec
        inc     r17
        cpi     r17,60
        breq    restart
        rjmp    top
```

## Passing Values to Subroutines

A subroutine, like the main program, has access to all the registers. Thus, any value left in the registers by the main program can be used or changed in the subroutine. In the following code, r16 and r17 have initial values from the main program. The subroutine changes both, but only r17's new value goes back to the main program due to the lack of a push and pull of r17.

```
        ldi    r16,3
        ldi    r17,4
        rcall sub
        ...
; — — — — — — — — — — — — — — — — — —
sub:    push   r16
        inc    r16
        inc    r17
        pop    r16
        ret
; — — — — — — — — — — — — — — — — — —
```

Variables are global and can be accessed anywhere in the program. In this code, the variable count is a 1-byte global variable initialized to 0. Being a global variable, the code is non-reentrant. The variable count can be accessed in both the main code and subroutine. The sensor subroutine reads port c and then if the value is non-zero, increments the variable count.

```
; count is a byte value at memory location 0x0100.
.equ    count = 0x0100
        ...
        ldi    r16,0
        sts    count,r16    ; make count zero.
        rcall sensor        ; check sensor
        ...
; — — — — — — — — — — — — — — — — — — — — — —
sensor: push   r16
        in     r16,pinc
        breq   sen1         ; skip if = 0.
        lds    r16,count
        inc    r16          ; increment count.
        sts    count,r16
sen1:   pop    r16
        ret
; — — — — — — — — — — — — — — — — — — — — — —
```

Values can also be passed to a subroutine via the stack. Keep in mind that after a value is placed onto the stack, the SP is decremented. The `rcall` instruction pushes the 2-byte return address onto the stack, and subsequently the SP is decreased by 2. The instruction `ret` does just the opposite. This next code passes 2 bytes to a subroutine via the stack. The main program puts the two values onto the stack followed by a 2, which is the number of numbers on the stack being passed. Next, the call to the subroutine puts the 2-byte return address on the stack. At this point, 5 bytes were placed on the stack. The subroutine transfers the SP value to the index register x so that x also points to the next available byte on the stack. The x register will be used to access the bytes passed on the stack. This list shows the stack values at this point.

| Stack | Comment |
| --- | --- |
| unknown | SP and x point to this |
| return addr. | Address of code to return to |
| return addr. | Address of code to return to |
| 2 | count, number of bytes passed |
| 33 | Second byte passed |
| 22 | First byte passed |

Three is added to register x in order to point to the count, or number of bytes passed on the stack. The count is put into r16, which is used to establish how many bytes to get from the stack. A loop is entered and count will be the number of times to loop. In the loop, each argument is pointed to by x, pulled off the stack into r17, and from there put into SRAM via index register y. Register y is initialized to 0x0000, which turns out to be the location for registers r0, r1, etc. The main purpose of this example is to illustrate how to pass values to a subroutine via the stack.

```
; ─────────────────────────
main:
    ldi     r16,22
    ldi     r17,33
    ldi     r20,2
    push    r16             ; put numbers onto stack
    push    r17
    push    r20             ; save a 2, the number of numbers to pass
```

```
        rcall    sub1
        pop      r20          ; pull numbers off stack
        pop      r17
        pop      r16
        ...
; ----------------------------
sub1:
        ldi      yl,0x00       ; y pointer points to memory location 0
        ldi      yh,0x00
        in       xl,spl        ; x pointer = stack pointer
        in       xh,sph
        adiw     x,3+0         ; x now points to last number on stack
        ld       r16,x+        ; get number of numbers passed
sub3:
        cpi      r16,0         ; done if 0 numbers to get
        breq     sub2
        ld       r17,x+        ; get number from stack
        st       y+,r17        ; save number to memory - or anywhere.
        dec      r16           ; decrement number of numbers remaining
        rjmp     sub3
sub2:
        nop
        ret
; ----------------------------
```

## AVR PERIPHERALS

The AVR includes digital ports, analog-to-digital conversion, interrupts, SPI, i2c, and other peripherals that are useful for communicating with the outside world. It is possible to connect LEDs, motors, lights, switches, potentiometers, GPS sensors, touch pads, temperature sensors, accelerometers, and other interesting devices via AVR's peripherals. For more information on any of these peripherals, refer to the manufacturer's data sheet. The ATtiny13 has 8 pins, the ATmega328 has 28 pins, and the ATmega32 has 40 pins. The ATmega32 has the most I/O capability. Be careful when connecting external signals to a microprocessor, as it is quite possible to damage it. Luckily, these microprocessors are inexpensive.

# Digital I/O Ports

The digital I/O ports are used in interfacing external hardware to the microprocessor. A signal can be sent out from the AVR to control a motor, or a signal from a sensor could be read into the AVR. Ports are either input or output as specified by a data direction register for that port. The ports usually comprise 8 bits each. Each port is mapped into memory, which means that each port has a memory address. The method for accessing a port is similar to that of a memory location. Note that an output port cannot source or sink much current—in fact, it generates just a few milliamps, which is enough to run an LED or transistor but never a motor. A transistor, in turn, could run a motor.

The ATmega32 has four ports: a, b, c, and d. The ATmega328 has three ports: b, c, and d. And the ATtiny13 has 6 bits of port b available. Often, each port is 8 bits. The direction of the port is determined by the data direction register (DDR) for that port. In other words, the DDRB bits determine if the PORTB pins are input or output. Furthermore, each bit of the port is individually controllable for direction. Upon reset, the AVR ports are set to input. For the port direction, a 0 makes the pin an input pin, while 1 makes the pin an output pin as shown by this example.

```
DDRB = 0b11000000;    //Makes pin  6 & 7 of portb output and the other pins input.
DDRD = 0b01000011;    //Makes pins 0,1,6 of portd output and the other pins input.
```

### ATmega328, Simplified View

Here is an overview of the ports used for digital input and output on the ATmega328.

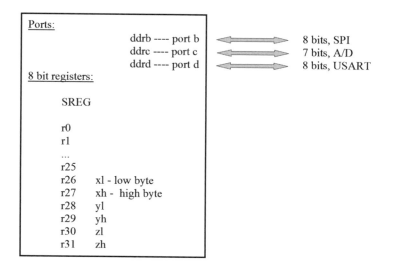

32 8-bit registers: r0 to r31

16-bit registers: x, y, z (mapped: r26 to r31)

DDRB, DDRC, and DDRD determines direction of ports b, c, and d.

It is important to restate that the ports are generally 8 bits, and each bit can be individually controlled. Some bits of a port could be input while others output. Each pin in the port is configured via the I/O registers. The port b registers are explained here. The AVR has other ports like A, C, or D, and they have similar registers to control those ports. The following three 8-bit registers control port b. PORTB and PINB both refer to the same port. The name PORTB is used when sending data out from the port, while PINB is used when inputting data to the port.

The DDRB (data direction register for portb) sets the direction of the bits for the port: 0 for input and 1 for output. The individual bits are named ddb0 to ddb7. Note that you must wait 1 cycle before using the port after changing the DDRB.

PORTB refers to the 8 output bits of port b. The individual bits are named portb0 to portb7.

PINB refers to the 8 input bits of port b. The individual bits are named pinb0 to pinb7. This is generally read-only, but writing a 1 to any bit will toggle the output bit of port b.

Before using the port, you must first set up the data direction register. To make all of port b input, set all the bits of DDRB to 0. For all the bits to be output, set all the bits of DDRB to 1. The 8 bits of DDRB can be any combination of input and output. Each bit of the port is individually controlled in terms of direction. Note that if you change the DDRB, then you must wait 1 cycle for the change to occur before using the port. You can use a nop instruction in the code after changing the DDR in order to wait the 1 cycle.

### Digital Output to a Port

A port like port b can be used as digital output (if the DDRB is set to ones). You can individually change any of the bits of the port. Remember to use the name PORTB when writing to a port. The instruction sbi can be used to set a single bit of a port, while cbi can be used to clear an individual bit, as shown by this section of code.

```
.equ   led = 2            ;Equates make code easier to read.
    sbi     portb,1        ;To set bit #1 of port b
    cbi     portc,led      ;To clear bit led.
```

These examples demonstrate different programming styles with regard to sending data out to port b in assembly. Don't forget to first set up the DDRB and then wait for 1 cycle for the DDRB to configure.

```
;Set up ddrb first.
    ldi    r16,0x0f      ; upper 4 bits are input, lower 4 bits are output.
    out    ddrb,r16      ; data direction register b
    nop                  ; Wait 1 cycle

;Example 1 - Set 2 bits using ldi.
    ldi    r16,0b11
    out    portb,r16     ;turn on bits 1 and 0

;Example 2 - Turns on bits using bit shift, <<
; | is a logical OR,
; & is a logical AND
    ldi    r16,(1<<pb2)|(1<<pb1) ;makes bits pb2 and pb1 a 1
    out    portb,r16                ;outputs r16 to portb

;Example 3 - Set bit using sbi, clear bit using cbi.
    sbi    portb,0                ;set   bit 0 of portb
    cbi    portb,1                ;clear bit 1 of portb

;Example 4 - Setting a bit in pinb actually toggles the portb bit.
    sbi    pinb,pinb1             ;toggles portb bit 1
    sbi    pinb,pinb2             ;toggles portb bit 2
```

Assembly is efficient, but often one desires the convenience of C. This next code shows how to program this in C. Note that in C you do not need to delay 1 nop after setting the DDRB due to the overhead C requires.

```
// Examples of working with portb in C.
// Here an LED is connected to port b bit 1.
  DDRB = 0b10;          // Bit 1 is output for LED, bit 0 is input
  PORTB = PORTB | 0b10; // | is a logical OR
  PORTB |=  0b10;       // Same as the line above.
  delay_ms(1000);       // Turn on LED, other bits not affected.
  PORTB &= ~0b10;       // & is a logical AND. ~ toggles all the bits.
  delay_ms(1000);       // Turn off LED, other bits not affected.
```

Notice how this next code is better than the prior code by using a name rather than a number. The #define replaces all occurrences of the first parameter with that of the second parameter before compiling occurs.

```
#define LED 0b10
DDRB |= LED;        // Bit 1 is output for LED, no other bits affected.
PORTB |= LED;       // Turn on LED, no other bits affected.
delay_ms(1000);
PORTB &= ~LED;      // Turn off LED, no other bits affected.
delay_ms(1000);
```

### Digital Input, Reading a Port

A port like port b can be used as digital input if the DDRB is set to zeros. You can read all 8 bits of port b and then mask out the bits that are not of interest. Remember to use the name PINB rather than PORTB when reading a port. The code is first written in assembly and then in C.

```
ldi     r16,0x00
out     ddrb,r16            ;all bits are input.
nop                         ;required delay for DDRB set up.
in      r16, pinb           ;read all 8 bits of port b
andi    r16, 0b00000010     ;isolates pinb1
```

Other useful assembly instructions you may want to use when reading a port include: andi, ori, cpi, brne, sbrs, sbrc, sbis, and sbic. The code in C will blink an LED while waiting for a button press, after which it will continue past the loop.

```
DDRB = 0b10;                // Bit 1 is output for LED, bit 0 is input
do {
    PORTB |=  0b10;
    delay_ms(200);          // LED on
    PORTB &= ~0b10;
    delay_ms(200);          // LED off
  } while (PINB & 0b01);    // Loop while switch, bit 0, is high.
```

### Input/Output Example 1

This code shows how to turn on four LEDs connected to the lower 4 bits of port b. Remember to first set the DDRB. The next bit of code will make port b an input port and then read all 8 bits, but then mask out the upper 4 bits, leaving the lower 4 bits unchanged.

```
; asm code to turn on the lower 4 bits of port b
    ldi     r16,0x0f
    out     DDRB,r16        ; set direction of port
    nop
    out     PORTB,r16       ; output to port.
```

```
; asm code to read the lower 4 bits of port b into r16
    ldi    r16,0x00
    out    DDRB,r16        ; set direction of port
    nop
    in     r16,PINB        ; read port
    andi   r16,0x0f        ; mask out upper 4 bits.
```

Here's the same problem solved in C:

```
// C code to turn on the lower 4 bits of port b.
    DDRB  = 0x0f;
    PORTB = 0x0f;
```

```
// C code to read the lower 4 bits of port b into x.
    DDRB = 0x00;
    int x = (PINB & 0x0f);        // read port and mask upper.
```

The last line in the preceding code shows how to mask out some of the bits while keeping other bits unchanged. The code (PINB & 0x0f) forces the upper 4 bits to zero while preserving the lower 4 bits. ANDing with a 1 will preserve the bit in question while ANDing with a 0 will force the bit to 0.

This diagram shows how to connect an LED and how to locate the + and − side of an LED. Some LEDs are round, while others may be square. The resistor value can be from 220 to 1k ohms or so. Lower values will make for a brighter LED but will draw more current.

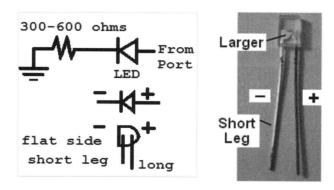

### Input/Output Example 2

Write a program to read a switch. The switch is connected to port b bit 1. When the switch is pushed and produces a low, turn on a motor. The motor is connected to port b bit 5 through an NPN switching transistor. Here's the solution with the circuit:

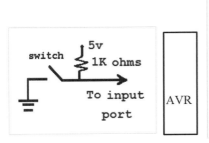
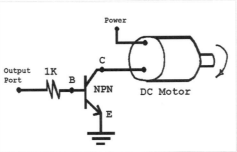

In assembly, the code is:

```
        ldi    r16,0b00100000
        out    DDRB,r16          ;set port direction
top:    in     r16,PINB          ;read port
        andi   r16,0b10          ;mask out all but bit 1
        brne   top               ;branch if = 0, keep looking

        ldi    r16,0b00100000
        out    PORTB,r16         ;turn on motor
```

In C, the code is:

```
DDRB = 0b00100000;
while ((PINB & 0b00000010) == 1);
PORTB = 0b00100000;
```

If port b is also being used for other devices, one should not disturb that configuration. The following code does that by only affecting the needed bits in the DDRB.

```
#define SWITCH 0b00000010
#define MOTOR  0b00100000
DDRB |= MOTOR;                    // Set motor pin as output
DDRB &= ~SWITCH;                  // Set switch pin as input
while ((PINB & SWITCH) == 1);     // Look for active switch
PORTB |= MOTOR;                   // Turn on motor.
```

### JTAG on '32

The JTAG interface is on some of the bits on port c of the ATmega32. It is used to debug a program as it runs. In order to use all of port c for I/O, turn off the JTAG. This code shows how to turn off the JTAG using in-line assembly. The reason for using assembly is that writing to MCUCSR must be done twice and within 4 cycles. C programming would be too slow. The pins of port c used for the JTAG are TMS, TCK, TDI, and TDO. If there is a chance that an interrupt could occur during this code, it would be a good idea to disallow interrupts for this section of code.

```
; Code to turn off JTAG for the ATmega32. ---------------------
; This code goes inside of a C program.
; Out must be done twice in 4 cycles.
asm (    "push  r18        \n"
         "in    r18,0x34   \n"        // MCUCSR=0x34
         "ori   r18,0x80   \n"
         "out   0x34,r18   \n"        // Turn off JTAG
         "out   0x34,r18   \n"        // Must be done twice.
         "pop   r18        \n");
      // Code to turn off JTAG. Must be done twice within 4 cycles. ----------
```

### Questions on Ports

Try answering these questions pertaining to ports.

1. Do the ports default to input or output when the machine is reset?

   Upon reset, the AVR ports are set to input.

2. What register controls the direction of port c?

   That would be the DDRC.

Make port c an input port. Then read port c until the lower 4 bits are all ones, without regard to what the upper 4 bits are.

```
int i;
DDRC = 0x00;
do { i = PINC;
     i = i & 0x0f;
    }while (i != 0x0f);
```

Make port c an input port with one switch connected to bit 0. Make port b output with 8 LEDs connected. Write the code that will read the value from port c. If bit 0 of port c is 1, then light all LEDs; otherwise, all LEDs are off. Loop forever.

```
DDRC = 0x00;
DDRB = 0xff;
while (1)
{   int i = PINC & 0b01;
    if (i==0b01) PORTB=0xff; else PORTB=0x00;
}
```

## Using a Comparator with a Sensor

Some sensors are analog in nature and need to be converted to either a high or low in order to be connected to a digital input port. The LM339 comparator is well suited to do this. To the output of the LM339, connect a 1k pull-up resistor to 5v. If the (−) input is higher than the (+) input, then the output is 0v; otherwise, the output is 5v. The set point at the (+) input can be produced using a potentiometer with 5v and 0v on the outside legs. The sensor in the following diagram produces a resistance that varies with what is being measured; thus, it needs another resistor effectively in series with it and having a value close to that of the sensor's average resistance. The output of the comparator can be connected to a digital input port.

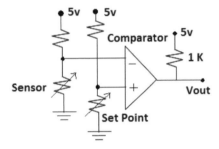

# Polling

Polling is a process in which the computer repeatedly checks a sensor for a value, and if the value is present, then the computer will run some code to do something. For example, the microprocessor could repeatedly read port b until bit 0 is a 1, and then when it is a 1 it will turn on a sprinkler system to put out a fire. When polling, the computer may not have much time left over to do other things like steer a car or execute pulse-width modulation (PWM).

In Example 1, below, port b bit 2 is connected to a sensor that goes low when active. When the sensor is active, the computer is to run a motor on port b bit 7 for 2 seconds. First, you need to set the direction of port b bit 7 as output by making the DDRB bit 7 a 1. The other bits are to be input, so you can make those bits of the DDRB a 0. PORTB is the name of the port when sending data out to the port, while PINB is the name of the same port when reading data. The instruction andi can be used to isolate a single bit like the one we wish to look at.

### Example 1: Polling a Sensor

```
.org    0x0000
    ldi    r16,0b10000000
    out    DDRB,r16
more:
    ldi    r16,0x00
    out    PORTB,r16          ;turn off motor
here:
    in     r16,PINB           ;read 8 bits of portb
    andi   r16,0b00000100     ;isolate pb2
    cpi    r16,0b00000100
    breq   here               ;keep looping while bit 2 = 1
    ldi    r16,0x80           ;turn on pb7
    out    PORTB,r16
    rcall  delay2s            ;a delay of 2 sec.
    rjmp   more               ;start again
```

### Example 2

The following section shows how to implement a fire detector using polling, which will work but is not as elegant as using an interrupt. With polling, the sensor does not go to an interrupt pin but rather to a digital input pin. As long as the sensor produces 5v and 0v, it can go directly to a digital input pin.

Fire Sensor Using Polling.

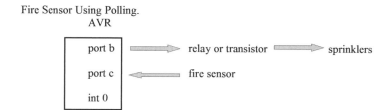

Here's the pseudocode for using polling:

Read port c bit for alarm.

If alarm active, turn on port b for sprinkler; otherwise, turn it off.

Loop forever.

On the other hand, if you have a sensor that produces 3v or more when active and less than 3v when not active, then using a digital input pin will not work. The reason for not sending this signal directly to a digital input pin is that while the 3v and above will produce a high, a voltage of 2v or more will also produce a high. Remember that any voltage above 2.0 is considered a high, while any voltage below 0.8 volts is considered a low. To interface this type of sensor to the AVR, you could use a comparator with a set point of 3v. Instead of using a comparator, another way to do the same thing is to use an analog-digital (A/D) input. At 3v, the A/D value would be $3/5 \times 255 = 153$. The A/D is discussed later in the chapter.

## Interrupts

Consider a phone on a desk in a room. If the ringer on the phone was not working, a person in the room would have to continually check the phone line to determine whether someone was calling. That's not very efficient, especially since most of the time no one is calling. This process is called *polling*. Or in the case of an intruder or fire, someone could check for this condition every few minutes. This polling process takes a lot of time away from someone or even a computer. If a ringer or alarm were added to the system, then the person or computer would not have to poll repeatedly for that special condition but could instead simply wait to be interrupted by the alarm when it occurs. This process is one of an *interrupt*. Interrupts are better suited when the condition is rare and unpredictable and the computer has other things to do (even sleeping). Polling is better when the sensor needs to be checked very quickly and the computer has nothing better to do.

Interrupts on the AVR allow for an external hardware signal to stop the current program and to jump to a service routine. An internal timer can also be used to trigger a periodic interrupt. The purpose of the service routine is to attend to the external condition. For example, a temperature sensor can be connected to the interrupt pin (INT0 or INT1) on the ATmega328. The INT0, INT1, and RESET pins on the ATmega328 are interrupt input pins. The reset pin is active-low, meaning that it activates when the voltage is low.

For the example shown in the following figure, when there is a fire the temperature sensor goes low. In response, the ATmega328 would stop its current task and then run the service routine that will turn on the sprinkler system (we hope!). After the fire is out and the service routine ends, the ATmega328 goes back to the process that was interrupted.

The fire sensor example could have been implemented using polling. In polling, the sensor is connected to one of the ports on the microprocessor, and the microprocessor would have to continually check the status of the sensor. This does not leave the microprocessor much time to do other tasks. The polling process generally takes up most, if not all, of the computing time. Imagine a large building. Someone or even a computer could spend all their time just polling each of the thousands of rooms for a fire.

AVR

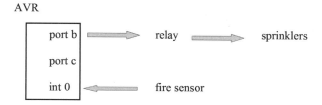

Simplified software:

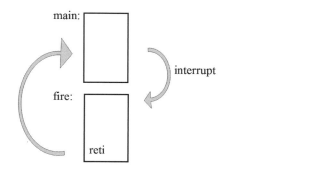

These are the detailed steps taken when an interrupt occurs:

1. The microprocessor is doing something (even sleeping).
2. The interrupt signal going to INT0 or INT1 becomes active.
3. The microprocessor finishes its current instruction.
4. If the I flag is 1, then the interrupt is allowed; otherwise, the interrupt is disallowed.
5. The program counter is pushed onto the stack.
6. The program counter (PC) is reloaded from the interrupt vector table (jump table). The PC now points to the service routine, which is the code to fix the issue.
7. The service routine runs until the instruction RETI is encountered.
8. The RETI pulls the PC return address from stack. The interrupted program is reinstated.

Note that the external interrupts INT0 and INT1 will trigger even if those pins are set as output. The interrupt can be triggered on a falling edge, rising edge, high level or a low level. A rising or falling edge requires the I/O pin not to be in a sleep mode. A low-level trigger can be used to wake up from a sleep mode.

The interrupt vector table is used to direct the AVR to the location of the service routine. It is somewhat confusing. The table holds an address where the service routine is to start. In the last step, the PC is reloaded from the stack in order to return to where it came from. The interrupt table is found in memory starting at 0x0000. The number of interrupts on the ATmega328 is 26. When the INT0 interrupt is active, the PC is reloaded with the address 0x0001. At this location, (0x0001) there needs to be a jmp instruction to the service routine. There are only 2 bytes (one instruction) available at 0x0001 for this jump. The programmer must put at 0x0001 a JMP to the service routine. In summary, upon seeing the interrupt, the AVR goes to the table, which takes it to the service routine. On the other hand, a reset interrupt will load the program counter with 0x0000. That is why there is a jmp to main in the examples. It is important to keep in mind that if the program does not use interrupts, then the program code is located starting at 0x0000 with no interrupt jmp instructions.

### Reset Interrupt

The reset button on the development board is connected to the reset pin and when pushed sends ground to the reset pin, which in turn resets the AVR. It is very similar to the reset button on a desktop PC. When a reset occurs, program execution will start at 0x0000 and the registers are set to their initial values. This interrupt never needs or has a reti instruction because it will never return to where it came from. It has the highest interrupt priority. Reset will occur when any of these occur:

- Power on
- Brown out
- External reset pin goes low
- Watchdog reset

### Watchdog

The purpose of the watchdog is to provide a mechanism for watching the AVR to make sure the program is not dead or hung up. It can be useful in situations such as when a space probe is traveling in space with no human to reset the computer if it were to crash. If the watchdog is activated, it requires the program to periodically load the watchdog counter register with a large number. The watchdog continually counts down toward zero. If the watchdog register counts down to zero before being reloaded with a large number, then the watchdog assumes the program is dead and will reset the program.

### Real Timer Interrupt

The real timer interrupt (RTI) can be used to generate interrupts at a fixed rate of time. This means one can have the program be interrupted every few milliseconds (or whatever time period) and run an interrupt subroutine. This RTI subroutine could be to check an input pin, steer a car, look for an object, blink a light, run a motor, etc.

### External Interrupt

An external interrupt, such as INT0 or INT1, is one of the pins on the AVR. To function as an interrupt, it would be connected to a sensor that detects a condition such as a fire. The sensor activates the pin, which in turn triggers an interrupt service routine (ISR) to fix the problem.

### Connecting Multiple Sensors to One External Interrupt

Most microprocessors have only a few external interrupt pins. If this is not enough, it is possible to connect several different sensors to one interrupt input through an OR gate. Any one of the sensors could activate the interrupt. Each sensor could represent a unique problem to service—one for fire, one for intruder, one for flood, and certainly one for pizza delivery, etc. How does the computer know which sensor is active and what to do? Since the interrupt pin does not know which sensor activated it, the computer will then have to poll each sensor. This may seem like a complicated polling system, but it's not really, because the polling is done only once after the interrupt is triggered.

Here the hardware is more involved and it does use more ports, but the computer does not have to continually poll the sensors. Using this method, the computer is free to do other things until an interrupt comes in. When the interrupt comes in, the first thing the service routine does is to read the input port (port c here). At least one sensor will be active. By looking at each bit of port c, one can determine the most important sensor(s) to attend to. In this circuit, the interrupt (IRQ) is low active, while the sensor shown is high active. How many sensors could the AVR easily work with using this method? Eight would be a good number.

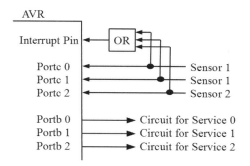

### Questions on Interrupts

1. With multiple sensors ORed into an interrupt, what instruction(s) can be used to find the active interrupt sensor? Assume that the sensors also go to port c.

The answer is to read port c, mask out the unused bits, and then the resulting value will pertain to the active sensor.

2. A ret instruction combined with what other instructions are equivalent to a reti? Hint: refer to the definition of reti.

Answer: The ret is used to end a subroutine, whereas a reti is used to end an interrupt. The ret just pops off the return address. The reti pops more bytes off a stack. To use the ret, one would have to determine what the interrupt pushed onto the stack and then pop them off.

### Interrupts on the ATtiny13

There are 10 different interrupts available on the '13 (see the following interrupt vector table list). If any of these interrupts occur, then the program counter will be reloaded with the appropriate address as shown in the table. This means that the program or interrupt service routine (ISR) will start to execute at one of these addresses. In the table, at the appropriate address, you must place a jump to the actual code or subroutine to run. (There is not enough room to put the ISR code in the table.) When the ISR is being executed, the hardware automatically disallows the current ISR to be interrupted (this can be overridden). As an example, if you want to use the watchdog, then put the instruction rjmp followed by the label in memory location 0x0008.

**ATtiny13 Interrupt Vector Table**

| # | Address | Source | Interrupt Definition |
|---|---------|--------|----------------------|
| 1 | 0x0000 | RESET | External Pin, Power-on, Brown-out, Watchdog |
| 2 | 0x0001 | INT0 | External Interrupt Request 0 |
| 3 | 0x0002 | PCINT0 | Pin Change Interrupt Request 0 |
| 4 | 0x0003 | TIM0_OVF | Timer/Counter Overflow |
| 5 | 0x0004 | EE_RDY | EEPROM Ready |
| 6 | 0x0005 | ANA_COMP | Analog Comparator |
| 7 | 0x0006 | TIM0_COMPA | Timer/Counter Compare Match A |
| 8 | 0x0007 | TIM0_COMPB | Timer/Counter Compare Match B |
| 9 | 0x0008 | WDT | Watchdog Time-out |
| 10 | 0x0009 | ADC | ADC Conversion Complete |

The External Interrupt Request 0 (pin INT0) is used to connect an external sensor as an interrupt to the AVR. The data direction register (DDR) does not need to be set for input for the external interrupt to work. The interrupt can be generated when the INT0 pin voltage does one of the four possibilities in the following table. The last three interrupts in the table require an active I/O clock to operate. To select one of these four possible trigger conditions, set bits isc01 and isc00 in register MCUCR appropriately.

| isc01 | isc00 | Interrupt is generated when int0: |
|---|---|---|
| 0 | 0 | is a low level |
| 0 | 1 | logically changes state |
| 1 | 0 | goes from high to low |
| 1 | 1 | goes from low to high |

The examples shown here use the low-level interrupt trigger condition. When this external pin goes low, the ATtiny13 will execute the rjmp located at 0x0001 and will return to where it came from when a reti instruction is encountered. The INT0 interrupt is pin 6 (pb1). To allow an INT0 interrupt:

1. Make isc01 and isc00 of mcucr (bit 1:0) a zero for low-level activation.

2. Set interrupt mask gimsk bit 6 (int0) to a 1.

3. Set I flag in sreg via instruction sei (cli will disallow int0).

The next example uses the timer interrupt. Place an rjmp instruction at memory location 0x0003. The label name chosen was "timer" for the interrupt service routine. Note that if you do not use interrupts, then the main program can begin right at 0x0000.

```
; ATtiny13 timer interrupt example.
.org       0x0000
   rjmp    main
.org       0x0003
   rjmp    timer
.org       0x000a        ; This org is not required.
main:
   ; Code to allow the interrupt source goes here.
   ; Your main code goes here.
```

```
timer:
    ; Your timer ISR goes here.
    reti        ; Return from interrupt.
```

The example shown next utilizes the external INT0 interrupt. The main program can always begin right after the last rjmp in the interrupt table, which saves a small amount of memory compared to the prior example.

```
; ATtiny13 int0 example.
.org    0x0000
    rjmp    main0        ; memory location 0x0000
    rjmp    int0_serv    ; memory location 0x0001
main0:
    ; Code to allow the interrupt source goes here.
    ; Your main code goes here.
int0_serv:
    ; Your timer subroutine goes here.
    reti        ; Return from interrupt.
```

The main program shown next will blink an LED five times a second. When an INT0 interrupt occurs, the ISR will stop blinking and turn on the LED for 4 seconds. Only the first org directive is really needed, as the second rjmp is at memory location 0x0001 even without the second org. Unfortunately, the instruction sbi gimsk,int0 does not work on the '13. Instead, it takes three lines to do the equivalent. Notice in the main program how writing a 1 to the appropriate bit in pinb toggles portb. The lines marked optional are not needed because on reset these bits are made zero.

```
; Example of ATtiny13 interrupts using the external int0.
; The main program blinks an LED.
; When an interrupt occurs, the service routine keeps the LED on for 4 seconds.
; To activate interrupt, ground the int0 pin.
.equ    led    = 2            ; led on portb2
.org    0x0000               ; reset
    rjmp    main0

.org    0x0001               ; int0
    rjmp    int0_service
main0:
    ldi    r16,(1<<led)       ; ddrd for led
    out    ddrb,r16

    ; ---------------- Code to allow int0 ----------------
    in     r16,mcucr          ; optional.
    andi   r16,~(1<<isc01)    ; optional. ;make these bits low
```

```
    andi   r16,~(1<<isc00)      ; optional.
    out    mcucr,r16            ; optional. ;Make int0 low level activation

    in     r16,gimsk
    ori    r16,(1<<int0)        ;enable global interrupts
    out    gimsk,r16            ;global interrupt mask

    sei                         ;allow interrupt int0
;  ───────────────────     Code to allow int0 ────────────────
main1:
    rcall  delay_100ms          ;fast blink and wait for interrupt
    sbi    pinb,led             ;toggle led
    rjmp   main1
; ────────────────
int0_service:                   ;Stop blinking. Turn on led for 4 sec.
    sbi    portb,led
    rcall  delay_1s
    rcall  delay_1s
    rcall  delay_1s
    rcall  delay_1s
    cbi    portb,led
    reti
; ────────────────
```

### Interrupts on the ATmega328

The '328 has more interrupts available than the '13. They work in the same fashion as on the '13. When the interrupt source becomes active, the program execution will jump to the associated address in the interrupt vector table. At the address in the table, there should be a jmp instruction to the interrupt service routine. The '328 supports both rjmp and jmp instructions, whereas the '13 supports only the rjmp instruction.

**ATmega328 Interrupt Vector Table**

| # | Address | Source | Interrupt Definition |
|---|---------|--------|----------------------|
| 1 | 0x0000 | RESET | External Pin, Power-On Reset, Brown-Out Reset, Watchdog System Reset |
| 2 | 0x0002 | INT0 | External Interrupt Request 0 |
| 3 | 0x0004 | INT1 | External Interrupt Request 1 |
| 4 | 0x0006 | PCINT0 | Pin Change Interrupt Request 0 |

*(Continued)*

**ATmega328 Interrupt Vector Table** *(Continued)*

| # | Address | Source | Interrupt Definition |
|---|---------|--------|----------------------|
| 5 | 0x0008 | PCINT1 | Pin Change Interrupt Request 1 |
| 6 | 0x000A | PCINT2 | Pin Change Interrupt Request 2 |
| 7 | 0x000C | WDT | Watchdog Time-Out Interrupt |
| 8 | 0x000E | TIMER2 COMPA | Timer/Counter2 Compare Match A |
| 9 | 0x0010 | TIMER2 COMPB | Timer/Counter2 Compare Match B |
| 10 | 0x0012 | TIMER2 OVF | Timer/Counter2 Overflow |
| 11 | 0x0014 | TIMER1 CAPT | Timer/Counter1 Capture Event |
| 12 | 0x0016 | TIMER1 COMPA | Timer/Counter1 Compare Match A |
| 13 | 0x0018 | TIMER1 COMPB | Timer/Counter1 Compare Match B |
| 14 | 0x001A | TIMER1 OVF | Timer/Counter1 Overflow |
| 15 | 0x001C | TIMER0 COMPA | Timer/Counter0 Compare Match A |
| 16 | 0x001E | TIMER0 COMPB | Timer/Counter0 Compare Match B |
| 17 | 0x0020 | TIMER0 OVF | Timer/Counter0 Overflow |
| 18 | 0x0022 | SPI, STC | SPI Serial Transfer Complete |
| 19 | 0x0024 | USART, RX | USART Rx Complete |
| 20 | 0x0026 | USART, UDRE | USART Data Register Empty |
| 21 | 0x0028 | USART, TX | USART Tx Complete |
| 22 | 0x002A | ADC | ADC Conversion Complete |
| 23 | 0x002C | EE Ready | EEPROM Ready |
| 24 | 0x002E | Analog comp | Analog Comparator |
| 25 | 0x0030 | TWI | Two-Wire Serial Interface |
| 26 | 0x0032 | SPM ready | Store Program Memory Ready |

To enable global interrupts on the '328, issue the instruction sei. Then set the bits in the register EIMSK to allow the appropriate interrupt. Remember that all interrupt service routines end with reti. The interrupt can be activated by a level change (high to low, or low to high), a toggle, or a low-level value on the interrupt pin. The bits isc in register EICRA determine how the interrupt is activated as shown by this table. The default is low-level value on the interrupt pin. No extra programming is needed for that condition.

| isc01 | isc00 | Interrupt is generated when int0: |
|---|---|---|
| 0 | 0 | is a low level |
| 0 | 1 | logically changes state |
| 1 | 0 | goes from high to low |
| 1 | 1 | goes from low to high |
| isc11 | isc10 | Interrupt is generated when int1: |
| 0 | 0 | is a low level |
| 0 | 1 | logically changes state |
| 1 | 0 | goes from high to low |
| 1 | 1 | goes from low to high |

This program uses the fact that the default value of isc01 is 0 and isc00 is 0. In this state, the ISR is activated when the interrupt signal is low. Note that this program demonstrates the use of two separate external interrupts at the same time. The jmp for INT0 and INT1 must be placed at memory locations 0x0001 and 0x0002, respectively. The jmp to main was placed at 0x0000, so the next instruction will, of course, be at 0x0001.

```
// ATmega328
.org    0x0000          ;IRQ jump table
        jmp     main
        jmp     int0_serv
        jmp     int1_serv

.org    0x0010
main:   sei                     ;allow interrupts
        sbi     EIMSK,INT0      ;allow int0
        sbi     EIMSK,INT1      ;allow int1
    ;... main program ...

int0_serv:
    ;... service routine 0...
        reti

int1_serv:
    ;... service routine 1...
        reti
```

### Interrupts on the ATmega32 in C

To enable interrupts in C on the '328, issue the instruction sei(). To enable the external interrupts INT0, INT1, or INT2, set those bits in the GICR register as shown by the example. The interrupt can be activated by a level change (high to low, or low to high), a toggle, or a low-level value on the interrupt pin. The bits isc in register MCUCR determine how the interrupt is activated, as shown by this table. The default is low-level value on the interrupt pin. No extra programming is needed for that condition.

| isc01 | isc00 | Interrupt is generated when int0: |
|-------|-------|-----------------------------------|
| 0 | 0 | is a low level |
| 0 | 1 | logically changes state |
| 1 | 0 | goes from high to low |
| 1 | 1 | goes from low to high |
| isc11 | isc10 | Interrupt is generated when int1: |
| 0 | 0 | is a low level |
| 0 | 1 | logically changes state |
| 1 | 0 | goes from high to low |
| 1 | 1 | goes from low to high |

The ATmega32 has the following interrupts.

**ATmega32 Interrupt Vector Table**

| # | Address | Source | Interrupt Definition |
|---|---------|--------|----------------------|
| 1 | 0x0000 | RESET | External Pin, Power-On Reset, Brown-Out Reset, Watchdog Reset, and JTAG AVR Reset |
| 2 | 0x0002 | INT0 | External Interrupt Request |
| 3 | 0x0004 | INT1 | External Interrupt Request 1 |
| 4 | 0x0006 | INT2 | External Interrupt Request 2 |
| 5 | 0x0008 | TIMER2 COMP | Timer/Counter2 Compare Match |
| 6 | 0x000A | TIMER2 OVF | Timer/Counter2 Overflow |

| 7 | 0x000C | TIMER1 CAPT | Timer/Counter1 Capture Event |
| 8 | 0x000E | TIMER1 COMPA | Timer/Counter1 Compare Match A |
| 9 | 0x0010 | TIMER1 COMPB | Timer/Counter1 Compare Match B |
| 10 | 0x0012 | TIMER1 OVF | Timer/Counter1 Overflow |
| 11 | 0x0014 | TIMER0 COMP | Timer/Counter0 Compare Match |
| 12 | 0x0016 | TIMER0 OVF | Timer/Counter0 Overflow |
| 13 | 0x0018 | SPI, STC | Serial Transfer Complete |
| 14 | 0x001A | USART, RXC | USART, Rx Complete |
| 15 | 0x001C | USART, UDRE | USART Data Register Empty |
| 16 | 0x001E | USART, TXC | USART, Tx Complete |
| 17 | 0x0020 | ADC | ADC Conversion Complete |
| 18 | 0x0022 | EE_RDY | EEPROM Ready |
| 19 | 0x0024 | ANA_COMP | Analog Comparator |
| 20 | 0x0026 | TWI | Two-Wire Serial Interface |
| 21 | 0x0028 | SPM_RDY | Store Program Memory Ready |

This example uses the external interrupt INT0 and the programming is done in C, which is a bit different compared to assembly. As in assembly, global interrupts and int0 have to be enabled. The data direction register (DDR) does not need to be set for input for the external interrupt to work. Do not forget to include interrupt.h as required for C programming. The name of the interrupt service routine is required to be ISR(INT0_vect). For other interrupt types, change the INT0 in INT0_vect to the appropriate interrupt name. As usual, the DDRC does not need to be set to input for the INT0 to work. The purpose of the main program is to toggle an LED on portb. The ISR will stop blinking the LED, keep the LED on for 2 seconds, and then return. Notice the global variable interrupt_count, which counts the number of times the interrupt was triggered. It needs to be global in scope so the interrupt service routine has access to it. It cannot be passed to the routine. It also must have the volatile modifier because the compiler has no idea when the ISR will activate and change the variable. This prevents the compiler from making assumptions as to the value of the variable. In this case, the interrupt_count can be used to know how many times the interrupt has been called so far. Here, if the number of interrupts is greater than 1000, then the delay is even longer.

```c
// ATmega32 int0 in C.
// Main program blinks an LED 4 times a sec.
// The interrupt stops the blinking for 2 sec.
#include <avr/io.h>
#include <avr/interrupt.h>
void delay_ms(int n);
volatile int interrupt_count = 0;
//-------------------------------
int main(void)
{ DDRB = 0xff;              // ddr for leds

// INT0 defaults to low level active.
// ISR(INT0_vect) runs when INT0 is active.
   sei();                   // enable global interrupts
   GICR |= (1<<INT0);       // enable int0 (bit 6) of gicr.

   while (1)
      {   PORTB = 0xff; delay_ms(50);     // Blink LEDs at this rate.
          PORTB = 0x00; delay_ms(200);
          // interrupt_count can be used here if need be,
      }
   return 0;
}
//-------------------------------
ISR(INT0_vect)                          // Stop blinking
{ PORTB = 0xff; delay_ms(2000);         // LEDs on for 2 sec.
  interrupt_count++;
  if ( interrupt_count > 1000 ) delay_ms(3000); // Increase delay
  return;
}
//-------------------------------
// n = the delay in milliseconds.
void delay_ms(int n)                    // Set for f = 1 MHz.
{   for (int t=0; t<n; t++)
      for (int i=0;i<54;i++);       // 1 msec.
}
//-------------------------------
```

## Analog to Digital Conversion

Analog to digital conversion, or A/D, is a technique for converting analog signals into digital values. Sensors that measure parameters such as temperature, angular position, pressure, and speed produce analog voltage signals on a single wire. An analog signal is a continuously variable voltage. A good example of this would be the graph of the outside temperature versus the time in hours. Computers store values in binary, typically using 8 or 16 bits. Analog signals are infinitely variable and can take on any value within their range. Binary values are discrete. A/D circuitry is used to make this conversion from the outside analog world to the digital world of the computer. In this process of converting from analog to digital, rounding errors occur, and in general this error is plus or minus one-half of the least significant bit.

A microcontroller like the AVR is often used to measure something in the environment. In the next table are some common parameters and a corresponding sensor. Sensors can vary in terms of size, cost, and performance. The sensor, which will be connected to the A/D converter, needs to produce a voltage that is linearly proportional to the parameter being measured. For example, if the speed doubles, then the speed sensor should double its voltage.

The AVR requires a reference voltage when using the A/D converter. This represents the maximum allowed voltage from the analog sensor. The reference voltage may not exceed Vcc. For example, a temperature sensor may produce voltages between 0v and 10v, depending on the temperature. In this case, the voltage should be scaled down to be within [5v,0v]. The LM35 temperature sensor is calibrated in degrees Celsius, is linear, and produces 10.0 mV/°C. If the temperature is 100 °C, then the output from the sensor is 1v. Here is a list of various sensors that produce analog signals.

| Parameter | Sensor/Part Number |
| --- | --- |
| angle | potentiometer |
| acceleration | accelerometer |
| humidity | Honeywell ih-3606 |
| light | photo resistor, IR pair |
| magnetic | HMC101 |
| pressure | Sensym ascx30an |
| speed | tachometer |
| | *(Continued)* |

| Parameter | Sensor/Part Number |
| --- | --- |
| strain | FlexiForce A401 |
| temperature | LM35 or thermal resistor |
| touch pad | tsr--51852 |

The A/D converter requires the signal going into it from the sensor to be properly conditioned. It requires the voltage from the sensor to be between 0v and the reference voltage, which is usually 5v. Anything outside that range can damage the A/D converter. For best sensitivity, the sensor should swing its voltage in the same range as the reference voltage. If the sensor voltage goes above the reference voltage, the sensor output can be reduced using a voltage divider. If the maximum sensor voltage is low, then the reference voltage can be reduced.

The concept of the A/D converter is simple. Given an 8-bit A/D converter with an input range of [0,5v] and the incoming voltage to the A/D converter is 2.5v, then the output is half of 8 bits or $255 \times (2.5/5) = 255/2 = 127 = 0x7f$. Notice the round-off error. The following table and graph show the relationship between the input voltage and the 8-bit equivalent using the reference voltage of 5v.

| Voltage | Conversion | Decimal | 8-Bit Value |
| --- | --- | --- | --- |
| 5v | 5/5 × 255 | 255 | 1111 1111 |
| 4v | 4/5 × 255 | 204 | 1100 1100 |
| 3v | 3/5 × 255 | 153 | 1001 1001 |
| 2.5v | 2.5/5 × 255 | 127 | 0111 1111 |
| 1v | 1/5 × 255 | 51 | 0011 0011 |
| 0v | 0/5 × 255 | 0 | 0000 0000 |

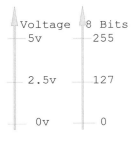

The circuit shown next shows how one could make a 4-bit D/A converter. It would be very easy to add more resistors in the same manner to get 8 bits. Notice the pattern of resistor values. Each bit input resistor has a ratio of 2 compared to the adjacent resistor.

The only challenge is getting the precise resistance. It looks good in theory; practice is another matter. The Rf value (2k feedback resistor) can be chosen to scale the maximum to the desired value. The negative can be taken out by another op-amp with a scaling of $-1$. A better D/A converter design is the R-2R circuit. Its resistors' values are more convenient. To derive the answer for Vout:

Assume the $(-)$ and $(+)$ inputs are the same voltage.

No current enters the $(-)$ or $(+)$ inputs.

Write a Kirchhoff's Current Law equation at the $(-)$ input.

A 4-bit D/A converter:

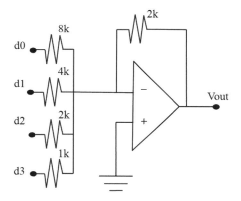

Vout = -2k( d3/1k + d2/2k + d1/4k + d0/8k )

A digital to analog (D/A) converter can be used to build an analog to digital (A/D) converter, as shown in the following figure. When reset occurs, the counter counts up. When the output of the D/A converter exceeds the analog signal, the comparator goes low (from a previous high), at which point the output of the AND gate goes low. The low from the AND gate stops the 8-bit counter. This results in the digital equivalent of the analog value. It could take as long as 255 clock periods to achieve the conversion. The op-amp in the circuit is configured as a comparator. If the $(+)$ signal is larger than the $(-)$ signal, the output is 5v; while if the $(-)$ signal is larger, the output is 0v.

An 8-bit A/D converter:

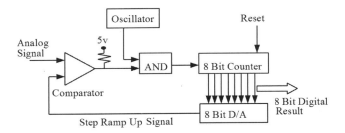

### Analog to Digital on the AVR

The A/D converter converts an analog voltage into a digital value that can be used by the processor. The range of voltages allowed is between 0 and Vref, where Vref should not be more than Vcc. Vcc is typically 5v or 3.3v. There are 4 A/D input channels on the ATtiny13, 6 on the ATmega328, and 8 on the ATmega32. The conversion takes anywhere from 13 to 260 microseconds depending on the analog voltage.

The A/D converter can take a single reading and stop or take non-stop readings. An interrupt can be used to start the A/D converter. In addition, when the A/D converter finishes it can generate an interrupt. These steps summarize what the programmer needs to do to use the A/D converter.

| | |
|---|---|
| 1. Connect the hardware | A/D channel and Vref |
| 2. Disable the digital input. | via register didr0 |
| 3. Select channel | via register admux |
| 4. Select 8- or 10-bit conversion | via register admux |
| 5. Start conversion | via register adcsra |
| 6. Enable A/D | via register adcsra |
| 7. Read result | via register adch and/or adcl |

The A/D converter (ADC) converts an analog input voltage to an 8- or 10-bit digital value. The lowest voltage allowed is 0v, and the maximum is Vref or the internal 1.1v reference voltage. In either case, do not allow the A/D input to exceed the reference voltage. Also, do not forget to connect Vref to the appropriate voltage. The reference voltage needs to be as stable and noise free as possible to ensure an accurate result. The AVR has several A/D channels. To select one, use the MUX bits in register ADMUX. The default channel is 0. To enable the ADC, set the ADC Enable bit, ADEN, in register ADCSRA. The ADC does not use power when ADEN is zero. Be sure to clear this bit before entering power-saving sleep modes.

The ADC generates a 10-bit result, which is found in the data registers ADCH and ADCL. By default, the result is presented right-adjusted (10 bits), but it can be presented left-adjusted (8 bits) by setting the ADLAR bit in register ADMUX. For an 8-bit result, simply read ADCH. For a 10-bit result, you must read ADCL first and then ADCH second. The ADC can trigger an interrupt when a conversion completes, or you can wait the appropriate time, which is at most 260 microseconds.

### A/D Using the ATmega328

There are 6 A/D channels (adc0 to adc5) on the '328, as shown by the following pin-out list. You must select the channel to read before reading a value from the channel.

| Pin Number | Pin Function |
| --- | --- |
| pin28 | pc5, adc5 |
| pin27 | pc4, adc4 |
| pin26 | pc3, adc3 |
| pin25 | pc2, adc2 |
| pin24 | pc1, adc1 |
| pin23 | pc0, adc0 |

The program needs to:

Select one of the 6 channels.

Select an 8- or 10-bit result.

Read continuously or read once.

In addition, do not forget to set Vref. Note that Vref is typically set to Vcc (or less). You must wait until the result is done (as long as 260 microseconds) or use the A/D Interrupt

Enable. To read A/D channel 0, 8 bits, non-stop, do all of the following shown in this outline:

1. Send the analog signal to pin23, A/D channel 0.

2. Wire 5v (Vcc) to pin21 (Vref).

3. Set didr0 = 0x3f for digital input disable.

4. Set admux = 0x20 for left-adjust, 8 bits, channel 0 select.

5. Set adcsra = 0xe0 to turn on A/D, start A/D, auto trigger.

6. Wait 260 microseconds or until ADSC = 0.

7. For 8-bit result, read adch.

8. For 10-bit result, read adcl and then read adch. Result = adch × 256 + adcl.

In assembly code form:

```
;Returns 8-bit A/D value in r16
    ldi    r16,0x3f        ;digital input disable.
    sts    didr0,r16
    ldi    r16,0x20        ;8 bits, select channel 0
    sts    admux,r16
    ldi    r16,0xe0        ;continuously update adch
    sts    adcsra,r16
    rcall  delay_1ms
    lds    r16,adch        ;Read A/D result into r16.
```

In C code form:

```
;Returns 8-bit A/D value in ADCH
    DIDR0  = 0x3f;          ;digital input disable.
    ADMUX  = 0x20;          ;8 bits, select channel 0
    ADCSRA = 0xe0;          ;continuously update adch
    while(ADSC==1);         ;wait for done
    int value = ADCH;       ;Read A/D result into value.
```

Now let's take a closer look at the registers associated with the A/D. The individual bits of these registers control various aspects of the A/D conversion process.

**ADMUX – ADC Multiplexer Selection Register**

| Bit | 7 | 6 | 5 | 4 | 3 | 2 | 1 | 0 |
|-----|---|---|---|---|---|---|---|---|
| Name | refs1 | refs0 | adlar | – | mux3 | mux2 | mux1 | mux0 |
| Initially | 0 | 0 | 0 | 0 | 0 | 0 | 0 | 0 |

Bits 7:6 – Ref, Voltage Reference Selection Bits

| Bits | Reference Selected |
|------|--------------------|
| 0 0 | External Aref, Default |
| 0 1 | External avcc |
| 1 1 | Internal 1.1v |

Bit 5 – ADLAR, ADC Left-Adjust Result

| | |
|---|---|
| 0 | Right-adjust, 10-bit result in adch:adcl |
| 1 | Left-adjust, upper 8-bit result in adch register |

Bits 3:0 – MUX, Analog Channel Selection Bits

| Bits | Channel Selected |
|------|------------------|
| 0000 | adc0, Default |
| 0001 | adc1 |
| 0010 | adc2 |
| 0011 | adc3 |
| 0100 | adc4 |
| 0101 | adc5 |
| 0110 | adc6 |
| 0111 | adc7 |

The result is in adch and adcl. If you want 8 bits only, make adlar=1 and read adch. For 10 bits, make adlar=0 and then you must read adcl before reading adch. Notice that the 10-bit result is adcl + 0x0100 × adch. This table illustrates this relationship.

```
If adlar=0 then   adch =   –     –     –     –     –     –   adc9 adc8
                  adcl = adc7 adc6 adc5 adc4 adc3 adc2 adc1 adc0
If adlar=1 then   adch = adc9 adc8 adc7 adc6 adc5 adc4 adc3 adc2
                  adcl = adc1 adc0   –     –     –     –     –     –
```

*(Continued)*

### ADCSRA – ADC Control and Status Register A

| Bit | 7 | 6 | 5 | 4 | 3 | 2 | 1 | 0 |
|---|---|---|---|---|---|---|---|---|
| Name | aden | adsc | adate | adif | adie | adps2 | adps1 | adps0 |
| Initially | 0 | 0 | 0 | 0 | 0 | 0 | 0 | 0 |

Bit 7 – ADEN, ADC Enable   1 = on 0 = off

Bit 6 – ADSC, ADC Start Conversion. 1 = start conversion(s). When done, it returns to 0.

Bit 5 – ADATE, ADC Auto Trigger Enable. 1 = auto triggering

Bit 4 – ADIF, ADC Interrupt Flag. Bit is set when an ADC conversion completes.

Bit 3 – ADIE, ADC Interrupt Enable. 1 = Activates A/D conversion complete interrupt if the I-bit in SREG is set.

Bits 2:0 – ADPS, ADC Prescaler Select Bits. Determines the division factor between the system clk and A/D clk.

### DIDR0 – Digital Input Disable Register 0

| Bit | 7 | 6 | 5 | 4 | 3 | 2 | 1 | 0 |
|---|---|---|---|---|---|---|---|---|
| Name | - | - | adc5d | adc4d | adc3d | adc2d | adc1d | adc0d |
| Initially | 0 | 0 | 0 | 0 | 0 | 0 | 0 | 0 |

Bits 5:0 – Digital Input Disable. 1 = Disable the digital input buffer on the corresponding ADC pin to reduce power consumption in the digital input buffer. One should do this when using the A/D.

### ADCSRB – ADC Control and Status Register B

| Bit | 7 | 6 | 5 | 4 | 3 | 2 | 1 | 0 |
|---|---|---|---|---|---|---|---|---|
| Name | - | acme | - | - | - | adts2 | adts1 | adts0 |
| Initially | 0 | 0 | 0 | 0 | 0 | 0 | 0 | 0 |

Bit 2:0 – ADTS, ADC Auto Trigger Source. If ADATE in ADCSRA = 1, then these bits select the trigger source (comparator, interrupt, timer). See data sheet.

Here is example C code that reads a 10-bit result by combining adch and adcl. Remember that adcl must be read before adch.

```
int b = adcl;
int a = adch;
int result = 256*a + b;
```

**A/D Example in C, 8 Bits, Using the '328**   For this example, connect a sensor that produces a voltage from 0 to 5v, to port c bit 0. Connect 5v to Vref of the '328. Here the software gets an 8-bit number from the A/D and puts it into the variable called value. More detailed information on the A/D converter for the AVR can be found in the data sheet from Atmel.

```c
// A/D, 8 bits, ATmega328 – – – – – – – – – – – – – – – –
#include <avr/io.h>
int ad_ch_init (int c);
void delay_1ms ();
int main (void)
{
  int value = ad_ch_init(0);
  // more code here to use value...
}
//– – – – – – – – – – – – – – – – – – – – – – – – – – – –
//Initializes A/D, selects channel c, and reads A/D, 8 bits
int ad_ch_init (int c)
{    DIDR0 = 0x3f;

     if (c==0) ADMUX = 0x20;      // pin 23
     if (c==1) ADMUX = 0x21;      // pin 24
     if (c==2) ADMUX = 0x22;      // pin 25
     if (c==3) ADMUX = 0x23;      // pin 26
     if (c==4) ADMUX = 0x24;      // pin 27
     if (c==5) ADMUX = 0x25;      // pin 28

     ADCSRA = 0xe0;           // Turn on A/D
     delay_ms(1);             // 1 millisecond delay
     return(ADCH);
} //– – – – – – – – – – – – – – – – – – – – – – – – – – –
```

**A/D Example in C, 10 Bits, Using the '328**   For this example, connect a sensor that produces a voltage from 0 to 5v, to port c bit 0. Connect 5v to Vref of the '328. This software gets a 10-bit number from the A/D and puts it into the variable value. Note that an integer variable can range from −32k to +32k, which will certainly hold a 10-bit value. When reading 10 bits, ADCL *must* be read before ADCH. To construct the 10-bit value, notice that ADCH is shifted left by 8 bits in the next to the last line. Another difference between this code and the 8-bit code is ADMUX bit 5. For a 10-bit result, bit 5 is a 0.

Reading 10 bits in assembly can certainly be done, but working with two 8-bit registers is not as easy as it is in C.

```c
// A/D, 10 bits, ATmega328 ----------------
#include <avr/io.h>
int ad10_ch_init (int c);
void delay_1ms ();
int main (void)
{
  int value = ad10_ch_init(0);
  // more code here to use the result, value.
}
//---------------------------------
// ADCH:ADCL form the 10-bit result but ADCL must be read first.
//Initializes A/D, selects channel c, and reads A/D, 10 bits
int ad10_ch_init (int c)
{    DIDR0 = 0x3f;

     if (c==0) ADMUX = 0x00;     // pin 23
     if (c==1) ADMUX = 0x01;     // pin 24
     if (c==2) ADMUX = 0x02;     // pin 25
     if (c==3) ADMUX = 0x03;     // pin 26
     if (c==4) ADMUX = 0x04;     // pin 27
     if (c==5) ADMUX = 0x05;     // pin 28

     ADCSRA = 0xe0;
     delay_ms(1);                // 1 msec delay

     int value = ADCL;           // lower 8 bits
     value += ADCH*0x0100;       // upper 2 bits added into value

     return(value);
} //---------------------------------
```

### A/D Using the ATtiny13

The A/D on the '13 works very much like that on the '328. One difference is that the '13 has fewer A/D channels. Also, the reference voltage on the '13 is either Vcc or the internal reference voltage. The registers used with the A/D are described next.

## ADMUX – ADC Multiplexer Selection Register

| Bit | 7 | 6 | 5 | 4 | 3 | 2 | 1 | 0 |
|-----|---|---|---|---|---|---|---|---|
| Name | - | refs0 | adlar | - | - | - | mux1 | mux0 |
| Initially | 0 | 0 | 0 | 0 | 0 | 0 | 0 | 0 |

Bit 6 – Reference voltage

0 = Reference is Vcc, Default

1 = Internal voltage reference

Bit 5 – ADLAR

0 = Right-adjust, 10 bits in adch:adcl

1 = Left-adjust, 8 bits in adch register

Bits 1:0 – Analog Channel Selection Bits

| Bits | Channel Selected | |
|------|------------------|---|
| 00 | adc0, pb5 | default |
| 01 | adc1, pb2 | |
| 10 | adc2, pb4 | |
| 11 | adc3, pb3 | |

The results are in adch and adcl. If you want 8 bits only, use adlar=1 and read adch. For 10 bits, make adlar=0, and then one must read adcl before reading adch.

| If adlar=0 then | adch = | - | - | - | - | - | - | adc9 | adc8 |
|-----------------|--------|---|---|---|---|---|---|------|------|
| | adcl = | adc7 | adc6 | adc5 | adc4 | adc3 | adc2 | adc1 | adc0 |
| If adlar=1 then | adch = | adc9 | adc8 | adc7 | adc6 | adc5 | adc4 | adc3 | adc2 |
| | adcl = | adc1 | adc0 | - | - | - | - | - | - |

## ADCSRA – ADC Control and Status Register A

| Bit | 7 | 6 | 5 | 4 | 3 | 2 | 1 | 0 |
|-----|---|---|---|---|---|---|---|---|
| Name | aden | adsc | adate | adif | adie | adps2 | adps1 | adps0 |
| Initially | 0 | 0 | 0 | 0 | 0 | 0 | 0 | 0 |

ADEN – A/D Enable. 1 = on, 0 = off.

ADSC – A/D Start Conversion. 1 = start conversion. 0 = conversion done.

ADATE – A/D Auto Trigger Enable. See data sheet for ATtiny13.

ADIF – A/D Interrupt Flag. The A/D Interrupt is executed if the ADIE bit and the I bit in SREG are set. The ADIF bit is cleared when the interrupt is being run, and then a 1 when the interrupt is finished.

*(Continued)*

ADIE – A/D Interrupt Enable. When set to 1 and when the I bit in SREG is 1, the A/D interrupt is enabled.

ADPS2:0 – A/D Clock Prescaler Select Bits. The system clock is scaled down before input to the A/D.

**DIDR0 – Digital Input Disable Register 0**

| Bit | 7 | 6 | 5 | 4 | 3 | 2 | 1 | 0 |
|---|---|---|---|---|---|---|---|---|
| Name | - | - | adc0d | adc2d | adc3d | adc1d | ain1d | ain0d |
| Initially | 0 | 0 | 0 | 0 | 0 | 0 | 0 | 0 |

Bit 5:2 – Digital Input Disable. 1 = disable the digital input buffer on the corresponding ADC pin to reduce power consumption in the digital input buffer. One should do this when using the A/D.

**Example 1**   This example is written in C and runs on the ATtiny13. It shows how to read channels 1 and 2. After selecting a channel, or between readings, one needs to wait up to 260 μsec. This program waits 1 msec. The 8-bit result is stored in variable s1 and s2 but is not used for anything in this program.

```
// Example A/D in C for the ATtiny13
#include <avr/io.h>
void delay_1ms ();

int main(void)
{
  DIDR0 = 0b10100;            // didr0 for pb2, pb4
  ADCSRA = 0xe0;              // continuously read A/D

  do{    ADMUX = 0b00100001;  //0x21;    // select channel 1
         delay_1ms();         // some delay is needed
         int s1 = ADCH;       // s1 = ch1 on pin 7, pb2

         ADMUX = 0b00100010;  //0x22;    // select channel 2
         delay_1ms();         // some delay is needed
         int s2 = ADCH;       // s2 = ch2 on pin 3, pb4
    } while (1);
  return (0);
}
void delay_1ms ()                   // Use 66 when f = 1.2 MHz.
{       for (int i=0;i<66;i++);     // Use 54 when f = 1 MHz.
}
```

**Example 2**  This code is for the ATtiny13. To use A/D channel 1, run the initialization subroutine, and then in the main program read the adch register for an 8-bit result. This example uses the A/D value to determine the position of a servo. The high time of a square wave going to a servo determines this position. The servo is on pb0 of the '13.

```
; ATtiny13_servo_ad.asm. Program will read A/D converter (ad1, pin 7)
;    and use that value to drive servo (pb0, pin 5).
; f = 1.18 Mcycles/sec. T = .85 usec/cycle
; ------------------------
.org    0x0000
    ldi    r16,0b01                ;pb0 is output to servo.
    out    ddrb,r16

    rcall  ATtiny13_ad1_init       ;initialize A/D
main1:
    ldi    r16,1                   ;send high to servo
    out    portb,r16
    rcall  delay_1p2ms             ;delay for 1.2 msec

    ; ------------------------
    ;Read A/D to add more time (0 to 635 usec.) to servo high time.
    ;Returns A/D value in r16 from adc1 (pin7, pb2).
    in     r16,adch                ;read result
main2:
    dec    r16                     ;each loop is 3 cycles or 2.5 usec.
    brne   main2
    ; ------------------------

    ldi    r16,0                   ;send low to servo
    out    portb,r16
    rcall  delay_18p5ms            ;delay 18.5 msec

    rjmp   main1
; ------------------------
; ------------------------
ATtiny13_ad1_init:       ;for A/D setup on pb2 of ATtiny13
    push   r16

    ldi    r16,0x3c
    out    didr0,r16                ;digital input disable
    ldi    r16,0x21
    out    admux,r16                ;8 bits channel 1
    ldi    r16,0xe0
    out    adcsra,r16               ;continuously update
```

```
    pop     r16
    ret
; — — — — — — — — — — — — — — — — — — — — — —
; — — — — — — — — — — — — — — — — — — — — — —
#include "attiny13_delays.asm"
; — — — — — — — — — — — — — — — — — — — — — —
```

## Serial Data Transmission

Transmitting data between devices or within devices can be done using different protocols. One method is to transmit the data in parallel, which means that there are many data wires (typically 8) plus the control lines connecting the devices. Another method is serial data transmission, which uses one data wire plus the control lines. The advantage of parallel transmission is a large bandwidth (fast data transmission). The advantage of the serial transmission is that one wire is cheaper than many wires, especially over long distances, or when there is limited space on the circuit board, or when chips have limited I/O pins. Reducing the number of I/O pins or wires can keep down the cost of production. Serial transmission can keep connectors smaller. This section will consider two types of serial data transmission, RS232 and SPI. RS232 is typically used between two devices, whereas SPI can be used to connect multiple devices together.

To understand how serial data transmission works, consider sending the message "Hello". Each letter is sent one at a time. First the "H", and lastly the "o". Each letter or character has a unique ASCII value (8 bits). It is this ASCII value that is transmitted. The ASCII value of the letter "H" is 48 hex or 01001000 binary. This number is sent out on the wire one bit at a time for a specific amount of time starting with the LSB or MSB, depending on the protocol selected. Certainly, both devices have to use the same protocol.

### RS232

RS232 serial data transmission is designed to connect two devices together. It utilizes one or two wires for data plus a ground. One data wire is used if the data goes only in one direction between the two devices. As an example, the letter "I" in ASCII is 49 hex or 01001001 binary. When this number is sent, it is sent out one bit at a time for a specific amount of time starting with the LSB and finishing with the MSB. Here the LSB is a 1 and the MSB is a 0.

With RS232 serial data transmission, more than just the ASCII bits are sent. In order to let the receiving device know when the data begins, the line is to idle high for sometime, then a start bit (low) is sent, then 8 data bits, and then 1 or 2 stop bits (high) are sent. The stop bits are the same as the idle high. The line is now ready for the next transmission.

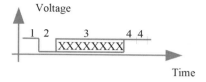

Each bit will be present on the wire for the same amount of time. The baud rate for serial transmission refers to the rate at which individual bits are sent. How long each bit is present on the wire is equal to 1 divided by the baud rate. For example, 9600 baud refers to 9600 bits per second, so each bit is present for 1/9600 of a second. The voltage out of the computer is often [0,5v], while the serial line voltage going into the target is [0,5v] or [−12v,12v]. A 1488 serial line driver chip can be used to convert between these voltages.

### RS232 Using the ATmega328 and ATmega32

The '328 and '32 have two pins available for RS232 serial data transmission. The voltages of these pins are 5v and 0v. The baud rate, parity, stop bits, and number of data bits are all set through register settings. After setting the registers, you can either send data or receive data via the UDR (uart data register). UBRR sets the baud rate and is actually two registers concatenated (UBRRH and UBRRL). Other registers used are the UCSRA, UCSRB, and UCSRC.

The default frequency of the '328 and '32 is 1 MHz. This can be increased to 8 MHz via the chip fuses so as to perform better at higher baud rates. For more information, see the chip's data sheet. Note that if you desire 12 volts for the serial line, then use the max232 chip between the serial line and the AVR. Pin 10 (data in) of the max232 is connected to pd1 (txd, transmit data pin) of the '328 or '32. Pin 9 (data out) of the max232 is connected to pd0 (rxd, receive data pin) of the '328 or '32.

| max232 | ATmega |
|---|---|
| pin 9, txd | pd0, rxd |
| pin 10, rxd | pd1, txd |

max232 pin-out:

- Pin 16 is Vcc, 5v.
- Place a 10 uf capacitor between pins 2 & 16, 4 & 5, 1 & 3, and 6 & 15.

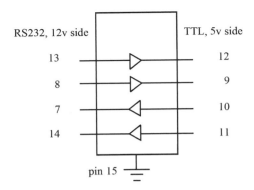

### RS232 Using the ATmega328 in Assembly

These three subroutines show, in assembly, how to initialize the USART0, how to transmit a byte, and how to receive a byte on the '328. The initialization sets up the protocol to be used. In this case, it is 4800 baud, 8 bits per byte, no parity, and 2 stop bits. These values are placed into the appropriate registers. The baud rate is converted into the number of clocks to tick by for each bit transmitted. The register ubrr0 represents the baud rate and is determined by:

- For double speed off, ubrr0h:ubrr0l = CPU clock frequency/(16 × Baud Rate) − 1
- For double speed on, ubrr0h:ubrr0l = CPU clock frequency/(8 × Baud Rate) − 1

The double clock speed is off in this example, which, if used, would double the number of clocks needed. The number of clocks is placed into UBRR0H:UBRR0L. The number of stop bits and the number of bits per byte is determined in the UCSR0C register. Transmit enable and receive enable is set in the UCSR0B register. To transmit a byte, first wait to see that the transmit buffer is empty, and then put the byte to transmit in register UDR0. To receive a byte, first wait to see that the MSB of UCSR0A is set to 1 and then read register UDR0.

```
; ─ ─ ─ ─ ─ ─ ─ ─ ─ ─ ─ ─ ─ ─ ─ ─ ─ ─ ─ ─ ─ ─ ─ ─ ─ ─ ─ ─ ─ ─ ─
;Assembly subroutines for ATmega328 using USART0
; ─ ─ ─ ─ ─ ─ ─ ─ ─ ─ ─ ─ ─ ─ ─ ─ ─ ─ ─ ─ ─ ─ ─ ─ ─ ─ ─ ─ ─ ─ ─
; Call this first. It initializes the USART0 to:
; 4800 baud, 8 bits, no parity, 2 stop bits
serial_init4800:
   push    r16

   ldi     r16,00
   sts     ubrr0h,r16    ; use sts instead of out.
                         ; double clock speed is off
                         ; ubrr =    6 for a baud rate of 9600
```

```
    ldi     r16,12        ; ubrr =   12 for a baud rate of 4800
    sts     ubrr0l,r16    ; ubrr = 207 for a baud rate of 300

    ldi     r16,(1<<txen0)|(1<<rxen0)      ;transmit/receive enable
    sts     ucsr0b,r16

    ldi     r16,(1<<usbs0)|(3<<ucsz00)     ;8 data, 2 stop
    sts     ucsr0c,r16

    pop     r16
    ret
; ---------------------------------
; ---------------------------------
;transmits r17 to the serial port via usart0
serial_t_r17:
    push    r16

serial_t1:
    lds     r16,ucsr0a     ;wait for empty transmit buffer. 'lds' not 'in'
    sbrs    r16,udre0
    rjmp    serial_t1
    sts     udr0,r17       ;put data in transmit buffer

    pop     r16
    ret
; ---------------------------------
; ---------------------------------
;Reads the serial port value into r17
serial_r:                  ; stop looping when ucsr0a.rxc0 bit set
    lds     r17,ucsr0a
    andi    r17,0x80       ; 0b10000000, rxc0 (bit 7) = 1 when done
    breq    serial_r

    lds     r17,udr0       ;read usart buffer.
    ret
; ---------------------------------
```

### USART0 Using the ATmega328 in C++

The USART (universal synchronous asynchronous receiver transmitter) uses the RS232 protocol. There is a USART on port d of the '328 and '32 that sends/receives serial data on the Tx/Rx pins, respectfully. You need to set up the appropriate registers and baud rate to initiate communication. The Tx goes to Rx on the other device and vice versa. In addition, a common ground is needed.

Here is an example program in C++ for the '328 demonstrating how to use the USART. First, the '328 sets up the USART to run at 9600 baud. Subsequently, the main program continually transmits the letter "c". The transmit pin must be connected to a device that that can receive it. A terminal window on the PC would work well for testing purposes.

```cpp
// Program to test serial data transmit in C++
#include <avr/io.h>
#include "serial_328cpp.h"              // See following code.
int main(void)
{   serial328 terminalwindow;           // Instance of serial328 class
    terminalwindow.init_tx(BAUD1200);   // Set to transmit at baud rate
    while(1) {   terminalwindow.tx('H');   // Send data
                 terminalwindow.tx('i');
            }
}
```

The completed subroutines are kept safe and out of sight in a separate file by using the include. This makes the main program more manageable, simpler, and apparently smaller. The file serial_328cpp.h is shown next. It contains the subroutines in C++ to initialize the USART0, to transmit, and to receive on the '328. Register UDR0 holds the value to be transmitted or received. A while loop is used to check to see if the transmit is ready or the receive is done.

```cpp
// serial_328cpp.h    Serial code in c++ for the atmega328.
// ---------------------------------------------------------
class serial328
{// Pin 3 (pd1) is TX. Pin 2 (pd0) is RX.
 // UBRR0 = n_clock = frequency/(8*Baud) -1 (Using double clock speed)
 //Baud Rate   vs. n_clock (Using double clock speed, f=1MHz)
 //9600            12      = frequency/(8*Baud) -1
 //4800            25      = frequency/(8*Baud) -1
 //2400            51      = frequency/(8*Baud) -1
 //1200            103     = frequency/(8*Baud) -1
private:
public:
        #define BAUD9600 12
        #define BAUD4800 25
        #define BAUD2400 51
        #define BAUD1200 103
        #define BAUD0600 207
  void init_rx (unsigned char n_clock); // Initialize USART to receive
  void init_tx (unsigned char n_clock); // Initialize USART to transmit
  void tx (unsigned char value);        // Transmit char
  char rx ();                           // Receive char
  void td (unsigned char x);            // Transmit char as a decimal
};
// ---------------------------------------------------------
```

```
// –––––––––––––––––––––––––––––––––––––––––––––––
void serial328::tx (unsigned char value)              // Transmit char
{ while ( (UCSR0A & (1<<UDRE0)) == 0 ); //(UDRE0==0);// Wait for transmit ready
  UDR0 = value;
  return;
}
// –––––––––––––––––––––––––––––––––––––––––––––––
// –––––––––––––––––––––––––––––––––––––––––––––––
char serial328::rx()                                  // Receive char
{ while ( (UCSR0A & (1<<RXC0)) == 0 ); // Wait for receive ready. RXC0=1 when done.
  char value = UDR0;
  return (value);
}
// –––––––––––––––––––––––––––––––––––––––––––––––
// –––––––––––––––––––––––––––––––––––––––––––––––
// Initialize USART to receive
void serial328::init_rx (unsigned char n_clock)
{ UBRR0H = 0;
  UBRR0L = n_clock;        // Set baud rate to ___ . See table.

  UCSR0A |= (1<<U2X0);     // 1=double clock speed on
//UCSR0B |= (1<<TXEN0);    // Transmit on
  UCSR0B |= (1<<RXEN0);    // Receive on

//UCSR0C default works also.
//UCSR0C |= (1<<UCSZ01) | (1<<UCSZ00) | (1<<USBS0); //8 bits, No parity, 2 stop
  return;
} // –––––––––––––––––––––––––––––––––––––––––––––
// –––––––––––––––––––––––––––––––––––––––––––––––
// Initialize USART to transmit
void serial328::init_tx (unsigned char n_clock)
{ UBRR0H = 0;
  UBRR0L = n_clock;        // Set baud rate to___ . See table.

  UCSR0A |= (1<<U2X0);     // 1=double clock speed on
  UCSR0B |= (1<<TXEN0);    // Transmit on
//UCSR0B |= (1<<RXEN0);    // Receive on

//UCSRC0 default works also.
//UCSR0C |= (1<<UCSZ01) | (1<<UCSZ00) | (1<<USBS0); //8 bits, No parity, 2 stop
  return;
} // –––––––––––––––––––––––––––––––––––––––––––––
// –––––––––––––––––––––––––––––––––––––––––––––––
// Transmits x in decimal. Transmits it as 3 ascii characters.
void serial328::td (unsigned char x)
```

```
{       if (x>=200) {tx('2'); x-= 200; }
   else if (x>=100) {tx('1'); x-= 100; }
   else              {tx('0');          }

        if (x>=90) {tx('9'); x-= 90;}
   else if (x>=80) {tx('8'); x-= 80;}
   else if (x>=70) {tx('7'); x-= 70;}
   else if (x>=60) {tx('6'); x-= 60;}
   else if (x>=50) {tx('5'); x-= 50;}
   else if (x>=40) {tx('4'); x-= 40;}
   else if (x>=30) {tx('3'); x-= 30;}
   else if (x>=20) {tx('2'); x-= 20;}
   else if (x>=10) {tx('1'); x-= 10;}
   else              {tx('0');        }

        if (x>=9 ) {tx('9'); x-= 9 ;}
   else if (x>=8 ) {tx('8'); x-= 8 ;}
   else if (x>=7 ) {tx('7'); x-= 7 ;}
   else if (x>=6 ) {tx('6'); x-= 6 ;}
   else if (x>=5 ) {tx('5'); x-= 5 ;}
   else if (x>=4 ) {tx('4'); x-= 4 ;}
   else if (x>=3 ) {tx('3'); x-= 3 ;}
   else if (x>=2 ) {tx('2'); x-= 2 ;}
   else if (x>=1 ) {tx('1'); x-= 1 ;}
   else              {tx('0');       }

   return;
}
// ---------------------------------
```

### USART0 Using the ATmega32 in C

The '32 has slightly different registers for serial data compared to the '328, but otherwise the steps are exactly the same. See the prior section on USART0 on '328 in C. This code does not contain a main program but rather only the three subroutines that demonstrate how to initialize, transmit, and receive data using the USART0.

```
// serial_32.c
// Code for ATmega32 and its serial port.
// This table shows how to select the n_clock value
//    given the clock speed and baud rate.
//                  baud:   300   1200   4800   9600
// 1MHz,single clock        207     51     12    5.5
//      double clock        416    103     25     12
// 8MHz,single clock       1666    416    103     51
//      double clock       3332    832    207    103
```

```
// Public subroutines available herein:
void serial_init (unsigned char n_clock);      // Call this to init.
                                               // n_clock=12 for 4800 baud and 1 MHz
void serial_t(unsigned char value);     // Pass to this a character to display.
unsigned char serial_r();          // Returns character arriving at serial port.
//-----------------------------------------------------------
// Waits for USART ready and then sends character called value.
void serial_t(unsigned char value)
{   while ( (UCSRA & (1<<UDRE)) == 0 );     // Wait for transmit ready
    UDR = value;
    return;
} //---------------------------------------------------------
//-----------------------------------------------------------
// Waits for, then receives character in USART and then returns char.
unsigned char serial_r()
{   // Wait for receive ready. RXC=1 when done.
    while ( (UCSRA & (1<<RXC)) == 0 );
    unsigned char value = UDR;
    return (value);
} //---------------------------------------------------------
//-----------------------------------------------------------
// This subroutine must be called first before serial_t or serial_r.
// It initializes the USART for use.
// For single clock, 1 MHz, 4800 baud, make n_clock = 12
void serial_init (unsigned char n_clock)
{
    UBRRH = 00;             // Set baud rate
    UBRRL = n_clock;        // Set baud rate to ___. See table.
    UCSRB |= (1<<TXEN);     // Transmit on
    UCSRB |= (1<<RXEN);     // Receive on
// UCSRC default is 0x86.
// Using 8 bits, No parity, 2 stop bits here.
    UCSRC |= (1<<URSEL) | (1<<UCSZ1) | (1<<UCSZ0) | (1<<USBS);
//   UCSRA |= (1<<U2X);     // 1=double clock speed, uncomment for double speed.
    return;
} //---------------------------------------------------------
```

# Serial Peripheral Interface (SPI)

SPI communication has an advantage in that many slaves can be connected to one master. For each slave added to the system, one more wire is required from the master. This is okay when the number of slaves is not too many. If the number of slaves becomes too great, you may want to consider using the I2C or Two-Wire Serial Interface protocol. It has a limit of 127 slaves while using just two wires. The software for the I2C is more complicated. In either case, one could design a system wherein one master talks to various sensors and LCD screens using these two protocols.

SPI can be implemented in software or in hardware. The '13, '328 and '32 each have a hardware implementation on port b. The software implementation is shown later, and it can be put on any port, but you must write the code to do this. The hardware SPI version is only on port b but on the upside can run in the background while other code is running. A SPI port typically has these signals:

| SPI Signal | Meaning |
|---|---|
| MISO | master in, slave out |
| MOSI | master out, slave in |
| /SS | slave select, one for each slave |
| SCK | slave clock |

Multiple devices can be connected together, but only one device will be set up as the master. The others will be configured as slaves. On each slave is a unique wire called slave select (/SS). It is controlled by the master, and it tells the slave when it is to be selected for communication. Only one of the slaves will be selected at any one time. To designate the master, connect its /SS to a high. All of the MOSI lines are connected together, all of the MISO lines are connected together, and all of the SCK lines are connected together.

To send a byte to a slave, the program initializes the SPI, selects the slave, and then places the byte to be sent in the master's SPDR register. The master's SCK will then start to clock the 8 bits of data out through the MOSI pin. The 8 bits in the master's data register will go to the slave, while at the same time 8 bits in the slave's data register will go to the master. This appears as a loop. It is not required that both data registers have pertinent data in them to send, but of course at least one of them will. The transfer is finished when the SCK completes 8 clocks. A similar process occurs when reading a byte from the slave. Again, the master must send a byte to the slave that might have no meaning to the slave, but the master will get a byte from the slave. This diagram shows the connections between one master and one slave.

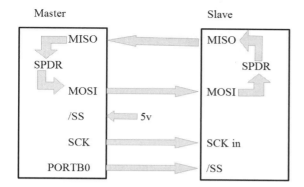

The next image shows how to connect several slaves to one master. Each slave gets the same GND, MOSI, MISO, and SCK from the master. Each slave gets a separate and unique slave select, /SS, from the master. In this example, the unused part of port b is used to select the slave.

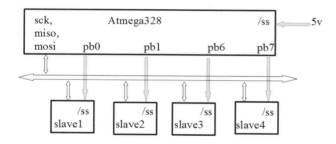

This program is an example of connecting two NHD0420D3Z LCD text screens to the SPI port of an ATmega328. Refer to the manufacturer's data sheet on this LCD for more information. Remember that the advantage of SPI is that it can easily connect multiple slaves to a master, as in this case.

```
/*
ATmega328_lcd_spi.c
SPI and LCD0420 Notes:
    spi slave may need to be removed to program avr.
    ss on master should be tied high when AVR program runs.
    ss on master should float when programming AVR.
    Programming cable does not need to be removed when AVR runs.
    Some LCD instructions need 2-4ms to complete.
    LCD brightness: 1=off, 8=max.
    Call spi_init() once for each device on SPI.
*/
#include <avr/io.h>
```

```c
#define LCD1    0b00000001    // ss    line for spi device. port b bit 0.
#define LCD2    0b00000010    // ss    line for spi device. port b bit 1.
#define MOSI    0b00100000    // mosi line for spi device.
#define SCK     0b00001000    // sck  line for spi device.

void spi_init(int device);
void spi_write(int device, char byte);
char spi_read(int device, char value);
void lcd0420_init(int lcd, int backlight);
void delay_ms(unsigned int n);

int main(void)
{
  spi_init(LCD1);
  spi_init(LCD2);
  lcd0420_init(LCD1,1);
  lcd0420_init(LCD2,1);

  for (int i='a';i<='z';i++)    // Display alphabet - - - - - - - - - -
    {
      spi_write(LCD1,i);            delay_ms(1);
      spi_write(LCD2,'z'-i+'a');  delay_ms(1);
    }
  while(1)    // Vary backlighting up and down. - - - - - - - - - - - - - - - - - - - - -
    {
    for (int i=1;i<9;i++ )
      { spi_write(LCD1,0xfe);
        spi_write(LCD1,0x53);
        spi_write(LCD1,i);
        delay_ms(100);
      }
    for (int i=9;i>0;i-- )
      { spi_write(LCD1,0xfe);
        spi_write(LCD1,0x53);
        spi_write(LCD1,i);
        delay_ms(100);
      }
    }
}
// Initialize spi - - - - - - - - - - - - - - - - - - - - - - - - - - - - - -
// The device is a byte with a 1 at the bit position that the device is located.
// Ex. If device is on PORTB3 then device = 0b00001000.
```

```
void spi_init(int device)
{
   DDRB |=    MOSI       //. mosi
            | SCK        //. sck
            | device     //. SS line for device

   // SPCR must match the protocol of the slave.
   SPCR =
            (1<<SPE)     //. enable spi
//        | (1<<DORD)    //. 1 = lsb first, 0 = msb first
          | (1<<MSTR)    //. master
          | (1<<CPOL)    //. 1 = sck idle high
          | (1<<CPHA)    //. 1 = sample sck on trailing edge (rising)
          | (1<<SPR1)    //. set clock rate fck/128
          | (1<<SPR0)    //.
            ;
}
//---------------------------------------
// Byte is the byte to be written to the device.
// The device is a byte with a 1 at the bit position that the device is located.
// Ex. If device is on PORTB3, then device = 0b00001000.
void spi_write(int device, char byte)
{
   PORTB &=~device;                // Select device
   SPDR = byte;                    // Load byte to data register
   while(!(SPSR & (1<<SPIF)));     // Wait for transmission complete
   PORTB |= device;                // Release device
}
//---------------------------------------
// To read an spi device, the master must transmit a byte to get a byte back.
char spi_read(int device, char value)
{
   PORTB &=~device;                // Select device
   SPDR = value;                   // Load byte to Data register
   while(!(SPSR & (1<<SPIF)));     // Wait for transmission complete
   value=SPDR;
   PORTB |= device;                // Release device
   return value;
}
//---------------------------------------
// Backlighting: 1=none, 8=bright.
void lcd0420_init(int lcd, int backlight)
```

```
{
  spi_write(lcd,0xfe);    spi_write(lcd,0x41);    delay_ms(4); // lcd on
  spi_write(lcd,0xfe);    spi_write(lcd,0x46);    delay_ms(4); // lcd home
  spi_write(lcd,0xfe);    spi_write(lcd,0x51);    delay_ms(4); // lcd clear
  spi_write(lcd,0xfe);    spi_write(lcd,0x53);                 // lcd backlight
  spi_write(lcd,backlight);                        delay_ms(4); // lcd backlight
  return;
}
//---------------------------------------------
// This delay was verified using the simulator.
void delay_ms(unsigned int n)
{
   for (int j=1;j<n;j++)
      for (int i=0;i<54;i++);   // This line takes 1 msec at 1 MHz.
}
```

## SPI Software Implementation

To implement SPI in software, on the master first choose a port with enough free bits—as a minimum, one bit for clock and one bit for MOSI. Then consider what is needed by the slave. If the slave can always be selected, then its select line can go directly to ground; otherwise, the master will need to have an output bit going to the slave select of the slave. If the slave is to return data to the master, the master will need to have an input bit for MISO. Don't forget that the master and slave will need to have a common ground between them.

This outline lists the steps, in software, to implement SPI. The MSB of the byte to transmit is sent first. Each bit is clocked on the rising edge. Once the hardware lines are connected between the master and slave, the master needs to:

1. Select slave by making the slave select bit a low.

2. Present each bit of the byte being sent via the following:

   Make clock bit low

   Send MSB of byte

   Make clock bit high, data captured when clock rises

   Delay if need be

   Repeat for the other 7 bits (LSB is sent last)

3. Deselect slave by making the slave select bit a high.

The master and slave must agree on the SPI protocol used. The aforementioned protocol is common but not set in stone. It is certainly possible that the master and slave send the LSB first or clock on the falling edge. The advantage of using a software implementation of SPI is that this method can use any part of any port. On the downside, the computer must spend its time sending out the bits rather than doing other things. In the upcoming example, C++ and classes are used. Classes tie together subroutines and data, thus reducing errors and making the code easier to write.

Let us summarize the rules of C++ and classes. A class is an idea or user-defined data type such as int, char, or float. An object of that class is just like declaring a variable of that type. The object of the class contains private and public members. The members of the class may be data like int, char, and float or may be code such as a subroutine. An object of the class can only call the public members of that class. The public members of a class can access the private members of that class. This prevents accidental access to the private data and subroutines. This section of code summarizes how a class is set up, in general. At first, it may look confusing, but in the end it makes for very good code. In the main program, notice the use of the class called lcd16x2. The object lcd is an instance of that class. In the last line of the main program, the method clear is executed on the object lcd. Note the use of the dot operator. The method called clear is essentially a subroutine, but it can run only on an instance of its class.

```
// Definition of class — — — — — — — — — — — — — — — — —
class lcd16x2
{    private:
         // subroutines or data go here.
     public:
         // subroutines or data go here.
         void clear()    { /* code for clear subroutine */ }
};
// main program — — — — — — — — — — — — — — — — — — — — —
int main(void)
{ class lcd16x2 lcd;   // lcd is the instance of the class lcd16x2.
  lcd.clear();         // Run the clear subroutine on lcd
}
```

Now look at the entire program. The program runs on an ATmega328. It reads the first 16 bytes from a GPS sensor and displays it on an LCD. It is possible to add more code to the main program that would parse the data and present latitude, longitude, speed, and direction. The GPS is connected to the '328 using the USART. The code to talk to the GPS is found in the file "serial_328cpp.h" and can be found in the section on the

USART0 for the ATmega328. The default GPS baud rate is 4800 and its data string starts with the character "$".

To communicate with the LCD screen, the program uses SPI implemented in software along with classes. The LCD is a Newhaven NHD-C0216CZ with two lines of 16 characters. It is very similar to how other LCDs operate. Refer to the class called lcd_16x2 shown next. It contains all the code needed to display characters on the LCD. The class contains six public members and three private members. The section of code that does the SPI implementation is found in the private member (subroutine) called out. The private method out does all the work of sending out the 8 bits plus controlling the clock and chip select lines. This subroutine is called by either the private subroutine dat if data is being sent to the LCD or by the private subroutine cmd if a command is being sent to the LCD. This particular LCD uses 3.3v and not 5v, while the GPS unit uses 5v. Having said this, it is easier to run the '328 on 3.3v as well. Connect the GPS power to 5v, and then its output data line will need to go to a voltage divider to reduce the voltage. The output data line goes to a 2k ohm resistor and then a 3k ohm resistor and then to ground. The USART's receive line of the '328 is connected to the point where the 2k and 3k meet.

```cpp
// This program takes the first 16 characters from a GPS sensor
//    and sends them to the LCD screen.
#include <avr/io.h>
#include "\avr c programs\lcd_16x2spi_cpp.h"
#include "\avr c programs\serial_328cpp.h"

int main(void)
{ class lcd16x2 lcd;
  class serial328 gps;

  gps.init_rx(25);   // initialize usart to 4800 baud
  lcd.clear();       // clear lcd

  do{
     do {} while ( gps.rx() != '$');  // Wait for start of string
     for (int i=0;i<16;i++)
       lcd.da( gps.rx() );            // Report the first 16 character
    } while (1);
}
//- - - - - - - - - - - - - - - - - - - - - - - - - - - - - - - - -
/* lcd_16x2spi_cpp.h       23 April 2014

atega328    LCD    Notes:
1       RST:   /reset. low to reset. Connect LCD reset to reset of AVR.
2       RS:    RS, Register Select. DI. 1=data,0=/instruction.
```

```
3        CSB:    /chip select. Tied to ground. Can be selected by AVR.
4        SCL:    spi clk
5        SI:     MO, mosi or Slave In.
6        GND
7        Vdd     3.3v

Note:    LCD uses 3.3v
*/
//---------------------------------------------
class lcd16x2
{
// LCD uses 3.3v! This does a software implementation of SPI.
   #define LCD16x2SPIS_PORT    PORTD
   #define LCD16x2SPIS_DDR     DDRD

// Bits in port for control lines
   #define LCD16x2SPIS_CS   0b00001000    //chip select
   #define LCD16x2SPIS_DI   0b00000100    //data,/instruction    same as:
   #define LCD16x2SPIS_SC   0b00000010    //spi clock
// #define LCD16x2SPIS_MO   0b00000001    // bit being sent, mosi. Same pin as usart rx
   #define LCD16x2SPIS_MO   0b10000000    //bit being sent, mosi. Frees up usart rx

    private:
       void dat (unsigned char data);    // Sends out data to display
       void cmd (unsigned char data);    // Sends a command to LCD.
       void out (unsigned char data);    // Sends out the bits using SPI

    public:
       lcd16x2();
       void clear()              {   cmd(01);    }
       void line1(int column)    {   cmd(0x80 | (0x00+column));}
       void line2(int column)    {   cmd(0x80 | (0x40+column));}
       void da(char c)           {   dat(c);     }
       void ds(char c[]);
};
//-----------------------------------------------------
//-----------------------------------------------------
// Constructor initializes LCD
lcd16x2::lcd16x2()
{
   LCD16x2SPIS_DDR   |= LCD16x2SPIS_CS;
   LCD16x2SPIS_DDR   |= LCD16x2SPIS_DI;
```

```
    LCD16x2SPIS_DDR   |= LCD16x2SPIS_MO;
    LCD16x2SPIS_DDR   |= LCD16x2SPIS_SC;

    // These are all needed to activate LCD
    cmd(0x30);              // display on
    cmd(0x39);              // function set
    cmd(0x14);              // internal freq.
    cmd(0x56);              // power control
    cmd(0x6d);              // follower control
    cmd(0x70);              // contrast
    cmd(0x0f);              // display on
    cmd(0x01);              // clear display
    for (int i=0;i<1000;i++);    // delay_ms(10);
}
//------------------------------------------------
//------------------------------------------------
// Displays the ascii string c ending with '$' on the LCD
void lcd16x2::ds(char c[])
{    int i=0;
    while (c[i] != '$')
        {    dat(c[i++]);
        }
}
//------------------------------------------------
//----------------------------
// Sends data to display on LCD. DI=0 for instruction, DI=1 for data
void lcd16x2::dat (unsigned char data)
{   LCD16x2SPIS_PORT |=  LCD16x2SPIS_DI;    // DI = 1;
    out(data);
}
//----------------------------
//----------------------------
// Sends a command to LCD. DI=0 for instruction, DI=1 for data
void lcd16x2::cmd (unsigned char data)
{   LCD16x2SPIS_PORT &= ~LCD16x2SPIS_DI;    // DI = 0;
    out(data);
}
//----------------------------
//----------------------------
// Uses a port and manually sends out the bits using SPI
```

```
// Rising edge of clk. MSB first.
void lcd16x2::out(unsigned char data)
{   LCD16x2SPIS_PORT &= ~LCD16x2SPIS_CS;            // CS = 0;
    for(int i=0; i<8; i++)
      { LCD16x2SPIS_PORT &= ~LCD16x2SPIS_SC;        // SCL = 0;
        if (data&0x80)
                LCD16x2SPIS_PORT |=  LCD16x2SPIS_MO; // Send a 1
          else  LCD16x2SPIS_PORT &= ~LCD16x2SPIS_MO; // Send a 0
        data<<=1;
        LCD16x2SPIS_PORT |=  LCD16x2SPIS_SC;        // SCL = 1;
        //delay if needed
        LCD16x2SPIS_PORT &= ~LCD16x2SPIS_SC;        // SCL = 0;
      }
    LCD16x2SPIS_PORT |=  LCD16x2SPIS_CS;            // CS = 1;
}
//----------------------------
```

## Clocks on the ATtiny13

This section describes clocks on the ATtiny13 and '85. The ATmega is very similar. See the data sheet from Atmel. There are four main clock signals on the AVR chip:

| Clock | Purpose |
|---|---|
| clk_adc | A/D converter |
| clk_io | General I/O modules |
| clk_cpu | CPU core and SRAM |
| clk_Flash | Controls the Flash and EEPROM memory |

The CPU clock is required to execute instructions in the program. The I/O clock is used by the timer/counter and some external interrupts. Halting the CPU and I/O clocks will improve the accuracy of the A/D converter. To reduce power consumption, clocks to the modules not being used may be halted. This is done using sleep modes. For the ATtiny13, the source of the clock can be one of these four sources. The default system clock for the ATtiny13 is 1.2 MHz, but remember that it is not exact.

| Clock Source | cksel fuses 1:0 | Note |
| --- | --- | --- |
| External clock | 0 0 | One can supply the clock to the AVR |
| Internal 4.8 MHz | 0 1 | |
| Internal 9.6 MHz | 1 0 | Default and divided by 8 yielding 1.2 MHz. |
| Internal 0.128 MHz | 1 1 | A low power clock |

The external clock frequency may range from 0 to 20 MHz at Vcc = 5v, but it may not change too quickly (2% maximum change from cycle to cycle). The maximum frequency drops off with decreasing Vcc. The clock prescaler is used to divide the clock source and is controlled by the Clock Prescaler Select, CLKPS3:0, found in register CLKPR. See the manufacturer's data sheet for more information. On the ATtiny13, the default clock source is the internal 9.6 MHz clock, but it is scaled down by 8 to get 1.2 MHz.

| CLKPS3:0 | CLK Division Factor | Note |
| --- | --- | --- |
| 0 0 0 0 | 1 | |
| 0 0 0 1 | 2 | |
| 0 0 1 0 | 4 | |
| 0 0 1 1 | 8 | Default setting |
| 0 1 0 0 | 16 | |
| 0 1 0 1 | 32 | |
| 0 1 1 0 | 64 | |
| 0 1 1 1 | 128 | |
| 1 0 0 0 | 256 | |

## Changing the Speed of the ATmega328

Using the internal clock source, the ATmega328 can run as fast as 8 MHz. It can also run slower to save energy when running on a battery. To change the clock speed, write the appropriate value to the CLKPR register found in the prior table. When changing speed, you must first disable interrupts and then set bit 7 of CLKPR to allow changes to CLKPR. These example subroutines show how to set the clock to 1 MHz and 8 MHz.

```
// 1 MHz – – – – – – – – – – – – – – – – – – – – – – – – – – –
void freq_1mhz ()
{       cli();
        CLKPR = 0B10000000; //enable CLKPR change
        CLKPR = 0B00000011; // set scale factor to 1
        sei();
}
// 8 MHz – – – – – – – – – – – – – – – – – – – – – – – – – – –

// 8 MHz – – – – – – – – – – – – – – – – – – – – – – – – – – –
void freq_8mhz ()
{       cli();
        CLKPR = 0B10000000; //enable CLKPR change
        CLKPR = 0B00000000; // set scale factor to 1
        sei();
}
// 8 MHz – – – – – – – – – – – – – – – – – – – – – – – – – – –
```

## Timer Counter

The '328 and '32 both have a 16-bit counter that can be used to time events. Also, the counter can be combined with a timer interrupt to run an ISR at regular intervals. When the counter is running, it starts at 0 and counts up to $2^{16}-1$, or 65,535, and then restarts. When it rolls over, the TOV1 flag is set. The rate at which it counts is $CLK_{I/O}$ divided by a prescaler. The prescaler can be set to 1, 8, 64, 256, or 1024. By default, the counter is turned off and you need to turn it on.

To turn on timer1 at the same speed as the $CLK_{I/O}$, make register TCCR1B = 0b001. This will make it count the fastest. If the clock is running at 1 MHz, then at this rate it will count from 0 to 65535 in 65,535 μsec or about 65 msec (1 clock cycle takes 1 μsec at 1 MHz). To turn it off, make TCCR1B = 0b000.

TCCR1B = 00000abc, where a, b, and c are the clock select bits.

| a b c | Clock select |
|-------|--------------|
| 0 0 0 | Counter stopped |
| 0 0 1 | clk io |
| 0 1 0 | clk io / 8 |
| 0 1 1 | clk io / 64 |
| 1 0 0 | clk io / 256 |

*(Continued)*

| a b c | Clock select |
|-------|--------------|
| 1 0 1 | clk io / 1024 |
| 1 1 0 | External clk on T1 pin, falling edge |
| 1 1 1 | External clk on T1 pin, rising edge |

TCNT1 is a 16-bit register that holds the timer count, although C programming does not show this. When TCNT1 is read, it is actually read in two 8-bit sections. This would be more readily seen in an assembly program over that of a C program. See the example subroutine shown later called `timer1_read()`. To prevent misreading or miswriting TCNT1, you must first disallow interrupts. This code example shows how to make a time measurement using the subroutines defined on the following pages.

```
timer1_on(1);
timer1_set(0);                          // zero the timer.
// some code or event occurs here that is to be timed.
unsigned int time = timer1_read();      // read timer.
```

Now the variable time has the time in microseconds, assuming the frequency = 1 MHz. Keep in mind that an unsigned integer ranges from 0 to 65,535, while integers range from −32K to +32K. So, for this example, the maximum time measurement would be 65,535 μsec.

Shown next is '32 code that can be used with the timer to make time measurements. First call `timer1_on()`. When turning on the timer, pass to it a scale value that corresponds to the clock rate desired. See the subroutines that follow. The `timer1_read()` subroutine will return the value of the timer count and thus how long it has been since the timer was turned on. This can be useful for measuring a timed event. The default I/O clock rate is 1 MHz. For more details on the timer, refer to the manufacturer's data sheet. This code can be used on the ATmega328, but first change register TIMSK to register TIMSK1.

```
// Timer code for the ATmega32.
// To use this on the ATmega328, change TIMSK to TIMSK1.
//----------------------------------
void timer1_off()
{    TCCR1B = 0b000;    // Turn off timer1,
}

//----------------------------------
// Turns on the timer. Scale determines the counting speed.
void timer1_on(int scale)
```

```
{
   if (scale==    1) TCCR1B=0b001; // timer1 on, count up, f=clk I/O rate/1
   if (scale==    8) TCCR1B=0b010; // timer1 on, count up, f=clk I/O rate/8
   if (scale==   64) TCCR1B=0b011; // timer1 on, count up, f=clk I/O rate/64
   if (scale== 256) TCCR1B=0b100; // timer1 on, count up, f=clk I/O rate/256
   if (scale==1024) TCCR1B=0b101; // timer1 on, count up, f=clk I/O rate/1024
}
// Read timer1 value. ──────────────────────
unsigned int timer1_read ()
{    unsigned char sreg = SREG; // Save global interrupt flag
   cli();                       // Disable interrupts
   unsigned int count = TCNT1; // Set i = TCNT1
   SREG = sreg;                 // Restore global interrupt flag
   return (count);
}
// Set timer1 to a value ─────────────────────
void timer1_set (unsigned int value)
{    unsigned char sreg = SREG;    // Save global interrupt flag
   cli();                 // Disable interrupts
   TCNT1 = value;         // Set TCNT1 to 0
   SREG = sreg;           // Restore global interrupt flag
}
// Enable the timer1 ISR ─────────────────────
void timer_isr_enable ()
{
   sei();                 // Enable global interrupts
   TIMSK = (1<<TOIE1); // Enable timer overflow int. Enable for ATmega32
// TIMSK1= (1<<TOIE1); // Enable timer overflow int. Enable for ATmega328
}
```

### Timer Counter with Interrupts

When the timer rolls over from 0xffff to zero, it can generate an interrupt that can be used to do some task (an ISR). First, allow global interrupts and the timer interrupt, and then turn on the timer. To allow interrupts, include this code in your main program:

```
#include <avr/interrupt.h>

   sei();                 // Enable global interrupts
   TIMSK1 = (1<<TOIE1);    // Enable timer overflow interrupt enable for '328
   TCCR1B = 0b001;         // Turn on timer1, count up, f = clk i/o rate/1.
```

Next, make an ISR that does some task. It must have this exact name: ISR (TIMER1_OVF_vect).

```
// Edit this to create your ISR, but do not change name of subroutine.
ISR(TIMER1_OVF_vect)
{ PORTB = ~PORTB;          // Example task, toggle LEDs
  return;
}
```

If using the slowest timer rate of CLK/1024 at 1 MHz, the ISR would trigger every 65.5 seconds. In some cases, this may not be long enough. To have a longer time between ISR events, you can use a global counter variable as shown in this code. In the example, after 1000 ISR events the ISR performs the task of toggling port b and then resets the counter to zero. The modifier volatile is needed, as the variable count may change unexpectedly and the compiler should not make assumptions about its value.

```
#include <avr/interrupt.h>

// Counts the number of times the interrupt triggers.
volatile unsigned int count=0;     // Must be global and volatile.

int main (void)
{ sei();                   // Enable global interrupts
  TIMSK1 = (1<<TOIE1);     // Enable timer overflow interrupt enable for '328.
  TCCR1B = 0b001;          // Turn on timer1, count up, f = clk i/o rate/1.

  // ... other code ...
}
//------------------------------
// Example of doing something once every 1000 interrupts.
ISR(TIMER1_OVF_vect)
{ count++;
  if (count==1000) {PORTB = ~PORTB; count=0;}
  return;
}
```

### Example of Playing a Piano Using Servos, Interrupts, and the Timer

This code plays a piano using eight servos. The servos need to strike each key as defined in the array called notes. A hammer connected to a servo is positioned up and down by a 20-msec periodic pulse waveform. The high time of the waveform is between 1.25 and 1.75 msec, which determines the position of the hammer. The timer interrupt is used to generate this waveform in the play_note() subroutine. The timer is set to time out whenever the waveform needs to change value. When the timer overflows and reaches 0, a timer interrupt is executed, which turns off the timer and says the variable done is true. Upon returning to the subroutine play_note(), the new value of the variable done, which is now true, kicks the program out of the infinite while(!done) loop. This variable, done, needs to be global because it cannot be passed to the ISR subroutine, and furthermore it needs to be volatile because otherwise the compiler will assume that the variable done is unchanged in the loop while (!done).

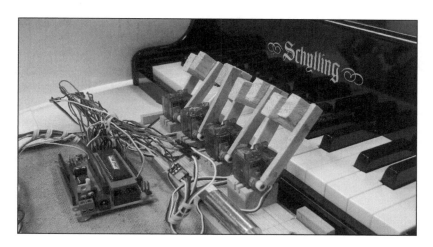

This next program plays the piano. It calls timer subroutines, which were shown just prior to this example.

```
//----------------------------------------------------------------
// ATmega32_piano.c
#include <avr/io.h>
#include <avr/interrupt.h>

#define port_servo        PORTD
#define port_servo_ddr    DDRD

            // Divide time (ms) by 20 msec to get # of servo cycles.
#define T_qn    300/20     // Time (ms) for 1 quarter note in servo cycles.
#define T_brk   120/20     // Time (ms) for 1 break between notes in servo cycles.
#define T    20000         // Period of servo in usec.
```

```c
#define T1    1250    // # usec for servo, down position
#define T2    1550    // # usec for servo,   up position

ISR(TIMER1_OVF_vect);
void play_note(unsigned char note);
void timer_isr_enable ();
void timer1_set (unsigned int value);
void timer1_off();
void timer1_on(int scale);

// The variable done is true (1) when the timer times out.
volatile int done = 0;    // Must be global and volatile.

//------------------------------
int main(void)
{
  #include "jingle_bells.c"    // File containing the notes.

  port_servo_ddr = 0xff;        // Set DDR
  timer1_off();                 // Timer off
  timer_isr_enable();

  while (1)
   {
     for (int n=0;n<nn;n++)
       {
         //............................. quarter note
         for (int i=0;i<T_qn;i++)
           play_note(note[n][0]);    // 1/50 sec for each play_note
         //............................. break between notes
         for (int i=0;i<T_brk;i++)
           play_note(note[n][0]);    // 1/50 sec for each play_note
       }
   }
  return 0;
}
//------------------------------

//------------------------------
// Produces the servo waveform.
// Line 1 makes a high for T1 time (the minimum high time).
// Line 2 completes the remaining high time to position servo.
// Line 3 makes a low for the remaining 20 msec.
void play_note(unsigned char note)
```

```
{
  port_servo = 0xff; timer1_set(0xffff-T1);
    done=0; timer1_on(1);
    while(!done);
  port_servo = note; timer1_set(0xffff-(T2-T1));
    done=0; timer1_on(1);
    while(!done);
  port_servo = 0x00; timer1_set(0xffff-(T-T2));
    done=0; timer1_on(1);
    while(!done);
}
//------------------------------
ISR(TIMER1_OVF_vect)
{ done = 1;          // used to end while(!done) loop.
  timer1_off();
  return;
}
// End of ATmega32_piano.c
//------------------------------------------------
```

This next file contains the notes and is included in the prior main program by using a #include.

```
//------------------------------------------------
// jingle_bells.c ---------------------------------
// These are the notes on the piano to strike. Stored in an array.
// The break is the rest time after the note. 0 = a break, 1 = no break.
int nn = 32;                      // number of notes
unsigned char notes[32][2] =
{
//    Notes          Breaks
//    cdefgabc        cdefgabc
   {0b00100000,    0b00000000,},    // jingle bells
   {0b00100000,    0b00000000,},
   {0b00100000,    0b00100000,},
   {0b00100000,    0b00000000,},

   {0b00100000,    0b00000000,},
   {0b00100000,    0b00000000,},
   {0b00100000,    0b00100000,},
   {0b00100000,    0b00000000,},
```

```
{0b00100000,    0b00000000,},
{0b00001000,    0b00000000,},
{0b10000000,    0b00000000,},
{0b01000000,    0b00000000,},
            // More notes not shown....
}
//- - - - - - - - - - - - - - - - - - - - - - - - - - - - - - - - - - - - - - - -
```

### Random Numbers

To generate a random number, you can use the timer and read the timer counter register TCNT0, which will be some unknown 8-bit value. First, turn on the counter via register TCCR0 (for the '32) or TCCR0B (for the '13 and '328) and then read TCNT0.

```
// Code to generate a random number
// TCCR0B |= 1;    // For '13 or '328. Turn on counter. No scaling.
   TCCR0  |= 1;    // For ATmega32.   Turn on counter. No scaling.
   // Wait an undetermined amount of time.
   // One could press a switch to get an undetermined amount of time.
   int x = TCNT0;  // Read counter, a random #
```

If the time between turning on the counter and reading TCNT0 is always the same, then the random number will always be the same, which is not good. To overcome this, you can wait for a button to be pressed after the counter is turned on. This will place an undetermined amount of time between turning it on and reading the counter, so the first number from the counter will not always be the same. This first random number can be used as a seed in this random number generator equation. The constants a, b, and m are picked to yield a sequence that does not readily repeat. The % means modulus, or remainder. The random number sequence will repeat after m numbers or sooner.

```
next random number = (last random number*a+b) % m
next random number = (last random number*106+1283)%6075 with a=106, b=1283, c=6075
```

The entire program in C is listed next. The values of x will range from 0 to 6074. If you want random numbers from 0 to 99, then use a statement such as: y = x%100.

```
#include <avr/io.h>
int main(void)
{
   TCCR0B |= 1;   // For '13, '328. Turn on counter. No scaling.
// TCCR0  |= 1;   // For ATmega32. Turn on counter. No scaling.
   // Wait for button pressed.
   // Thus x is not always the same.
```

```
DDRB = 0x00;
while(PINB & 0b1);

int x = TCNT0;     // Get first random number
// Generate 255 pseudo random numbers
for (int i=0;i<255;i++)
  {   x = (x*106+1283) % 6075;     // with a=106, b=1283, c=6075
  }
}
```

# Waveform Generation

The AVR microprocessors can easily produce periodic waveforms on their output pins by looping and sending out highs and lows. The drawback is that the AVR would spend nearly all of its time generating the waveform and leaving very little time to do other things. The solution to this is to use the AVR's built-in waveform generation. Once set up, the waveform is automatically generated, and the AVR is then completely free to do other things. Periodic waveforms are often used to blink LEDs and run DC motors. To control the speed of a DC motor, use pulse-width modulation (PWM).

### Square Wave Generation Using the '328

The ATmega328 can be used to generate a square wave with 50% duty cycle and with a range of frequencies. Once the program sets up the registers to produce a square wave, the program can go on to do other things while the square wave is generated in the background. Note that this would not work well for controlling the speed of a motor, as the duty cycle is fixed at 50%. For controlling the speed of a motor, use PWM. For more information on square wave generation, see Atmel's AVR data sheets.

The two square-wave outputs of the '328 are found on these pins:

| Pin | Function |
| --- | --- |
| pin 11 | oc0b, pd5 |
| pin 12 | oc0a, pd6 |

The frequency of the square wave is determined by:

$$f_{oc0a} = f_{clk\_io}/(2N(1+ocr0a))$$

$$f_{oc0b} = f_{clk\_io}/(2N(1+ocr0b))$$

where N is 1, 8, 64, 256, or 1024, as determined by tccr0b.

There are two square-wave outputs called oc0a and oc0b. They can be individually turned on with different frequencies. To generate a square wave, the counter needs to be turned on and scaled. The Output Compare Register (ocr0a or ocr0b) needs to be loaded with a number. The Clear Timer on Compare Match (CTC) mode needs to be selected along with the toggle mode by updating register tccr0a. When the counter counts up to and matches the Output Compare Register, the counter will be cleared to 0. At the same time, the output pin (oc0a or oc0b) will toggle, and the process repeats.

The Output Compare Register A, ocr0a, is used to determine frequency for pd6.

The Output Compare Register B, ocr0b, is used to determine frequency for pd5.

Choose one of these settings to select the output (in the toggle mode and CTC mode)

tccr0a = 0100 0010 = 0x42 For oc0a

tccr0a = 0001 0010 = 0x12 For oc0b

tccr0a = 0101 0010 = 0x52 For oc0a and oc0b

tccr0b = 00000abc where a, b, and c are the clock select bits.

| a b c | Clock Select |
|-------|--------------|
| 0 0 0 | counter and waveform stopped |
| 0 0 1 | clk io / 1 |
| 0 1 0 | clk io / 8 |
| 0 1 1 | clk io / 64 |
| 1 0 0 | clk io / 256 |
| 1 0 1 | clk io / 1024 |
| 1 1 0 | external clk on T0 pin, falling edge |
| 1 1 1 | external clk on T0 pin, rising edge |

Note that the appropriate port d pins must be set to output using register DDRD.

DDRD = _ 1 1 _ _ _ _ _ sets pd5 and pd6 to output since:

oc0b is on pd5

oc0a is on pd6

Next is an example program to blink an LED at the rate of 1.9 Hz. Using N = 1024 and OCR0A = 255, the equation for frequency then yields 1.9 Hz. An LED is connected to pd6 (pin 12). Once the square wave is set up, it will run automatically in the background and the program is then free to do other code.

```
#include <avr/io.h>
int main(void)
{
  DDRD    = 0x40;     // Sets Data Direction Register D, bit 6.
  TCCR0A = 0x42;      // Toggle & CTC mode.
  TCCR0B = 0x05;      // Rate = CLK/1024.
  OCR0A  = 0xff;      // Used to determine frequency. Frequency is 1.9 Hz.
                      // Square wave is now running.
  while (1);
}
```

### PWM Using the ATmega328, 8 Bit

There are two pulse-width modulation (PWM) outputs. They can have different duty cycles but will have the same frequency. In addition, there are two different types of PWM. One is fast PWM and the other is Phase Correct (PC) PWM. The two PWM output pins are:

| Pin | Function |
| --- | --- |
| pin 11 | oc0b, pd5 |
| pin 12 | oc0a, pd6 |

The frequency of the PC PWM wave is given by: $f = f_{clk\_IO}/(N*510)$.

The frequency of the fast PWM wave is given by: $f = f_{clk\_IO}/(N*256)$.

where N is 1, 8, 64, 256, or 1024.

The 1-byte registers ocr0a and ocr0b determine the duty cycle. A value near 0x00 will produce a duty cycle near 0, whereas a value near 0xff will produce a duty cycle near 100%. The frequency of the fast mode is twice that of the phase-correct mode for the same value of N. See the product data sheet for more information. To set up PWM, set the registers according to the tables.

To select mode, use tccr0a, Timer Counter Control Register A.

```
tccr0a  = 1 0 _ _ _ _ 0 1 = 81 hex    ;PC   PWM mode for oc0a
tccr0a  = 1 0 _ _ _ _ 1 1 = 83 hex    ;fast PWM mode for oc0a
tccr0a  = _ _ 1 0 _ _ 0 1 = 21 hex    ;PC   PWM mode for oc0b
tccr0a  = _ _ 1 0 _ _ 1 1 = 23 hex    ;fast PWM mode for oc0b
```

To select frequency, use tccr0b, Timer Counter Control Register B. Both PWM outputs will have the same frequency.

tccr0b = 00000abc where a, b, and c are the clock select bits for both outputs.

| a b c | Clock Select | Notes |
| --- | --- | --- |
| 0 0 0 | counter stopped | |
| 0 0 1 | clk io | 4 kHz for fast PWM, 2 kHz for PC PWM. |
| 0 1 0 | clk io / 8 | |
| 0 1 1 | clk io / 64 | |
| 1 0 0 | clk io / 256 | |
| 1 0 1 | clk io / 1024 | 3.8 Hz for fast PWM, 1.9 Hz for PC PWM |
| 1 1 0 | external clk on TO pin | falling edge |
| 1 1 1 | external clk on TO pin | rising edge |

Registers ocr0a and ocr0b are the output compare registers for oc0a and oc0b, respectively. They are used to set the duty cycle of the periodic wave. The duty cycle is the time the wave is high divided by the period. To set the duty cycle to 10%, let ocr0a be 10% of the maximum, 255. This turns out to be 25 base 10. Note that port d must be set to output using DDRD.

Here is an ATmega328 example of blinking an LED at 1.9 Hz and 25% duty cycle on pin 12. The maximum value for ocr0a is 0xff, and 25% of that is 0x40. Connect an LED to pin 12.

```
ddrd    = _ 1 0 _ _ _ _ _ = 40 hex   ; data direction register d.
tccr0a  = 1 0 _ _ _ _ 0 1 = 81 hex   ; PC PWM for oc0a
tccr0b  = 0 0 0 0 0 1 0 1 = 05 hex   ; freq.
ocr0a   = 40 hex                     ; for duty cycle of 25%
```

In assembly code form:

```
ldi    r16,0x40
out    ddrb,r16         ;ddrd
ldi    r16,0x81
out    tccr0a,r16       ;PC PWM mode for oc0a
ldi    r16,0x05
out    tccr0b,r16       ;2 Hz freq.
ldi    r16,0x40         ;0x40/FF is 25%
out    ocr0a,r16        ;duty cycle, 25% on pin 12
```

In C code form:

```
DDRB       = 0x40;
TCCR0A     = 0x81;
TCCR0B     = 0x05;
OCR0A      = 0x40;
```

### PWM Using the ATtiny13 or '85, 8 Bit

PWM on the ATtiny13 and ATtiny85 is the same as the ATmega328, with the exception of the port used for PWM. The '13 and '85 use pb0 and pb1 instead of pd5 and pd6. Otherwise, the code is the same.

| Pin   | Function   |
|-------|------------|
| pin 5 | ocr0a, pb0 |
| pin 6 | ocr0b, pb1 |

This ATtiny85 or ATtiny13 program is an example of fast PWM with a frequency at about 4 kHz. A 25% duty cycle waveform is on ocr0a, while a 75% duty cycle waveform is on ocr0b.

In assembly code:

```
ldi    r16,0x03
out    ddrb,r16         ;ddrb

ldi    r16,0xa3
out    tccr0a,r16       ;a & b fast PWM mode
ldi    r16,0x01
out    tccr0b,r16       ;4 kHz freq.
```

```
ldi     r16,0x40
out     ocr0a,r16           ;duty cycle, 25% on pin 5 (ocr0a)
ldi     r16,0xc0
out     ocr0b,r16           ;duty cycle, 75% on pin 6 (ocr0b)
```

In C code:

```
DDRB        = 0x03;     // Data Direction Register
TCCR0A      = 0xa3;     // A & B fast PWM mode
TCCR0B      = 0x01;     // 4 kHz
OCR0A       = 0x40;     // 25% on ocr0a
OCR0B       = 0xc0;     // 75% on ocr0b
```

## EEPROM on the ATmega328 in C

The EEPROM memory of the AVR can be used to store information even when the power is turned off. Very often, it is used to store data that is acquired. If power is lost, the data is still retained, compared to SRAM, which will lose its values when the power is off. To upload the EEPROM data to your PC, run AVR Studio. Select Tools > AVR Programming. Type in the device name and click Apply. Click on Memories and then click Read.

This example, in C, uses subroutines that read from and write to the EEPROM. The subroutines are found in <avr/eeprom.h>. One must include that header file. To write your own subroutines, refer to the AVR data sheet on the EEPROM.

```
eeprom_write_byte ((uint8_t *) address_to_write_to, char value_to_write);
char value_read = eeprom_read_byte((uint8_t *) address_to_read_from);
```

Example program of reading from and writing to the EEPROM:

```
//    atmega328_eeprom1.c
//    The NHD16x2 LCD uses 3.3v !!!!!!!!!!!!!!!!!

#include <avr/io.h>
#include <avr/eeprom.h>
#include "\avr\avr c programs lock box\lcd_nhd16x2spi.h"

int main(void)
{
    lcd16x2spi_init();                    // Initialize LCD for use
// Send a message ending in a $ to the EEPROM
    char m2[] = "Hi dad $";
    i=0;                                          // i = address to write to
    do   { eeprom_write_byte ((uint8_t *)i, m2[i]);     // write to EEPROM
         } while (m2[i++]!='$');
```

```
// Read a message from the EEPROM ending in a $
   char EEByte;
   i = 0;                                        // i = address to read from
   do   { EEByte = eeprom_read_byte((uint8_t *)i++);    // read from EEPROM
          lcd16x2spi_da(EEByte);                 // display data on LCD
          }while (EEByte!='$') ;
   while (1);
}
```

## PROGRAMMING LANGUAGE SUMMARY

Using AVR Studio, the AVR microprocessor can be programmed in assembly, C, C++, or a combination thereof. C has the advantage of convenience, whereas assembly is fast and efficient. This section contains example programs of each language. The last topic, in-line assembly, is interesting in that it combines assembly in a C program to have the advantages of both languages.

### Programming in Assembly

The following example program in assembly shows how to construct a main program and subroutines. The subroutines illustrate how to read a single bit in a port and how to make a delay. The main program demonstrates how to set the data direction register (DDR) of a port, how to set a bit in a port, how to loop, and how to call a subroutine. The purpose of the program is to slowly dim an LED. Here the LED turns on when given a low.

```
; ─────────────────────
.equ     led = 5              ;LED is on port c bit 5 and is low active.
.org     0x0000
;Sets direction of port c.
;Port c pins 5,4 are output via the 1s.
;All other pins are inputs due to the 0s.
     ldi     r16,0b00110000
     out     ddrc,r16
;Sets direction of port d.
;All pins of port d is made input via the 0s
     ldi     r16,0x00
     out     ddrd,r16
;This section will slowly dim an led ────────────
main0:
     ldi     r16,0xff            ;initial on time of led
     ldi     r17,0x01            ;initial off time of led
```

```
main4:
    cbi     portc,led           ;A zero turns on led. For r16 * 50 usec
    push    r16
main2:
    rcall   del_50us
    dec     r16
    brne    main2
    pop     r16

    sbi     portc,led           ;A 1 turns off led. For r17 * 50 usec
    push    r17
main3:
    rcall   del_50us
    dec     r17
    brne    main3
    pop     r17

    inc     r17                 ;increase off time by 50 usec
    dec     r16                 ;decrease on time by 50 usec
    brne    main4
;End of the section that dims an led  ----------------

    rjmp    main0               ;restart
;  ----------------------
;  ----------------------
; Read pinb bit 1 until it is a 1.
wait_pb1h:
    push  r16

wait_pb1h1:
    in      r16,pinb
    andi    r16,0b10
    breq    wait_pb1h1

    pop     r16
    ret
;  ----------------------
;  ----------------------
; Read pinb bit 1 until it is a 0.
wait_pb1l:
    push  r16

wait_pb1l1:
    in      r16,pinb
    andi    r16,0b10
    brne    wait_pb1l1
```

```
        pop     r16
        ret
; ─ ─ ─ ─ ─ ─ ─ ─ ─ ─ ─ ─ ─ ─ ─ ─ ─ ─
; ─ ─ ─ ─ ─ ─ ─ ─ ─ ─ ─ ─ ─ ─ ─ ─ ─ ─
; Delay 10ms or so. 10ms = 200 loops of 50us.
del_10ms:
        push    r16

        ldi     r16,200
del_10ms1:
        rcall del_50us
        dec     r16
        brne    del_10ms1
        pop     r16
        ret
; ─ ─ ─ ─ ─ ─ ─ ─ ─ ─ ─ ─ ─ ─ ─ ─ ─ ─
; ─ ─ ─ ─ ─ ─ ─ ─ ─ ─ ─ ─ ─ ─ ─ ─ ─ ─
; Delay 50us or so. 4 cycles/loop * 12 loops = 48 cycles = 48us.
; 1 cycles = 1 usec at 1 MHz.
; The overhead time should also be included.
del_50us:
        push    r16

        ldi     r16,12
del_50us1:
        nop
        dec     r16
        brne    del_50us1

        pop     r16
        ret
; ─ ─ ─ ─ ─ ─ ─ ─ ─ ─ ─ ─ ─ ─ ─ ─ ─ ─
; ─ ─ ─ ─ ─ ─ ─ ─ ─ ─ ─ ─ ─ ─ ─ ─ ─ ─
msg1: .db   "hello mom"    ;Demonstrates how to include data in a program.
; ─ ─ ─ ─ ─ ─ ─ ─ ─ ─ ─ ─ ─ ─ ─ ─ ─ ─
```

## Programming in C

For the most part, C is easier to program in, but it does not execute as fast as assembly does and the code size tends to be larger. This first program example illustrates the syntax of C and statements such as bitwise operators, ports, looping, and subroutines. Note the use of the built-in delays from AVR.

```c
//Examples of C syntax.
#define F_CPU 1000000UL     // Needed for the AVR delays. Freq = 1.0 MHz.
#include <avr/io.h>
#include <util/delay.h>     // Needed for the AVR delays

void step (short bit_pattern, double t);    // Subroutine prototypes

int main(void)
{ DDRD = 0b0011;        // Data direction registers
  DDRC = 0xf0;          // 1=output bit, 0=input bit

// Bit wise operators: ----------------
//    &, bit and
//    |, bit or
//    ~, bit not.

// Example of looping until a button is pressed -----
 short buttons;
 do { buttons = PINC;       // read all 8 bits
      buttons &= 0b01;      // isolate bit 0.
    } while (buttons);      // loop while bit 0 = 1.

// Example of a do while() loop. -----------
do {   step (0b11000000,50);
       step (0b01100000,50);
       step (0b00110000,50);
       step (0b10010000,50);
    } while (1);            // This loops forever.

// Example of a for loop ----------------
   for (int i=1;i<28;i++)
     { PORTD = i;           // Send i to portd
       _delay_ms(500);
     }

// Example of a while() loop ------------
   int i=55;
   while(i>0)
       { i--;           // Decrement i
         PORTD = i;
       }

// Example of a do while() loop ----------
   i=8;
   do{ PORTD = i--;         // PORTD equals i, then i is decremented.
      _delay_ms(1000);
     } while (i);           // Same as while(i != 0)
```

```
// Example of looping forever ----------------
   while (1);
}
//-----------------------------------------
//-----------------------------------------
// Example subroutine. Outputs bit_pattern to the stepper motor port.
// Two values are passed in. The t represents the time in ms.
void step (short bit_pattern, double t)
{ PORTC = bit_pattern;
  _delay_ms(t);
}
//-----------------------------------------
```

The following second example of C programming shows how to make a delay subroutine. You can use the AVR simulator to verify that the delay is for 1 ms. The #define replaces the first item with the second item throughout the code. Don't forget that subroutines need prototypes. This code reads all 8 bits of portb (remember to use pinb for input), and if any one of them is a 1, then the delay is run.

```
#include <avr/io.h>
#define SWITCH        PINB
#define SWITCH_DDR    DDRB
#define LEDS          PORTD
#define LEDS_DDR      DDRD

void delay_ms (int t);

int main(void)
{
    SWITCH_DDR = 0x00;        // Set data direction register to input
    LEDS_DDR   = 0xff;        // Set data direction register to output
    if (SWITCH & 0b01) delay_ms(2000);    // Wait if pinb0 is high
    LEDS = 0xff;              // Light LEDS
}

//-----------------------------------------
void delay_ms (int t)        // t is the # of msec to delay for.
{
  for (int j=0;j<t;j++)
     for (short i=0;i<54;i++);    // This loop is about 1 ms
  return;
}
//-----------------------------------------
```

## Programming in C++

C++ allows the use of objects and classes. The advantage of a class is that all the related data and subroutines are all tied together. For example, all the code and data for a particular sensor can be tied together into a class. This helps prevent data or subroutines from being misused, and it also reminds the programmer of the relationship between the subroutines and data.

Take the example of an LCD screen connected to a '328. Traditionally, you would write subroutines for the LCD, such as this:

```
void lcd16x2_init();
void lcd16x2_clear();
void lcd16x2_display_char(char c);
void lcd16x2_display_decimal(int n);
```

How do you know that these four subroutines all work together on one device? One does not know, except that they have similar names! Now consider this code implemented using a class. As you can see, the subroutines and any data are bound together in the class.

```
class lcd16x2
{   private:
        int column, row;
    public:
        lcd16x2();
        void clear();
        void display_char(char c);
        void display_decimal(int n);
}
void lcd16x2::lcd16x2()
{ // code for this subroutine goes here, such as:
  Column = 0; row = 0;
}
```

Inside the class, list the subroutine prototypes, and then outside the class write the definition of the subroutine. Items in the private section can be accessed only from within the public section. Only the items in the public section can be accessed from outside the class. A subroutine with the same name as the class is called a constructor. For the prior code, lcd16x2() is the constructor. It is run when a variable is made of that class. The constructor can be passed information but may not have a return type or value. The main program would look like the following. Notice the use of the dot notation.

```
int main (void)
{
    class lcd16x2 lcd1, lcd2;    // makes 2 variables of this class
    lcd2.display_char('H');      // displays an H on lcd2.
}
```

## In-Line Assembly

C is very convenient to program in, but sometimes assembly is desired for increased speed and smaller code size. It is possible to put assembly code inside of C. On entry (and exit), it is recommended to push (and pop) all registers used as shown next. For example, this C program uses assembly to output a high on all pins of PORTB.

```
// Within a C program ...
DDRB = 0xff;
asm (   "push    r18          \n"
        "ldi     r18,0xff      \n"
        "out     PORTB,r18     \n"
        "pop     r18           \n"
     );
// Now back to C.
```

This next example turns off the JTAG interface on the '32 so that one can use all of port c; otherwise, some of the pins are unusable.

```
// Within a C program ...
asm ( "push    r18          \n"
      "in      r18,0x34      \n"      // MCUCSR=0x34
      "ori     r18,0x80      \n"
      "out     0x34,r18      \n"      // Turn off JTAG
      "out     0x34,r18      \n"      // Must be done twice.
      "pop     r18           \n"
    );
// Now back to C.
```

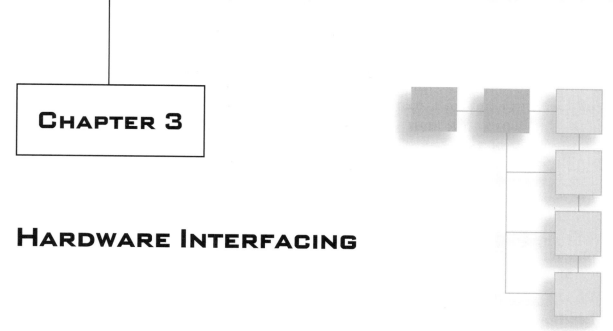

# CHAPTER 3

# HARDWARE INTERFACING

This chapter concentrates on how to interface various hardware devices such as sensors, motors, and displays to an AVR. Some hardware is easier to interface than others are. Before buying hardware, you should read the manufacturer's data sheet to see what is entailed to communicate with it. For example, a sensor with an A/D connection is easier to interface than an SPI device, which is easier than an I2C device. On the other hand, SPI and I2C allow many devices to be connected at once. Some devices can take up a lot of I/O on an AVR and thus not leave room for other devices to be connected. The I/O requirements will help determine which AVR is best suited for the project. Some hardware devices, like motors, take a lot of power and may require their own power supply.

You may want to develop and finalize code for various devices and then save that code in some sort of lockbox folder for future use on other projects. With this lockbox of debugged code, you don't have to redevelop the same code each time.

## SENSORS

Sensors are used to bring information into the AVR. There is a wide range of sensors, what they measure, and their communication protocols. In a project, try to use sensors that are easy to work with, or at least do not debug multiple new sensors at one time. This will make debugging easier and thus reduce the time required to complete the task.

## Digital Switch

A switch can be used to turn something on. It can also be used to select a high or low voltage, which can be sent into an AVR port. This allows the user to tell the AVR what to do. Here is a switch circuit that will produce 5v or 0v depending on the position of the switch. With the switch closed, the output is set to ground, 0v. With the switch open and virtually no current flowing through the resistor, the output voltage is very nearly 5v.

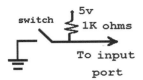

If the value of the pull-up resistor is too large, the current will cause a significant voltage drop from the 5v. That voltage drop is subtracted from the 5v, which may result in the voltage from the switch no longer being a high. On the other hand, if the pull-up resistor is too small, then when the switch is closed the current will be larger than desired. 1K ohms works well. This C code will read the switch until it is pressed (assume that the switch is connected to port b bit 0).

```
DDRB = 0x00;
while (PINB & 0b01);
```

## Reed Switch

A reed switch is a basic two-legged switch. It closes in the presence of a magnetic field. There is no switch lever that someone could toggle. A reed switch could be used, for example, to detect the limit of motion for a robotic arm. When the robotic arm, with a magnet on it, moves over a reed switch, the switch will close. If the switch is connected to a digital input port on the AVR, the AVR can sense this and in turn stop the arm.

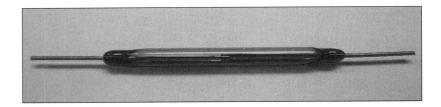

## Transistor as a Switch

The transistor can act as an electrical switch to turn something on, such as a motor or light. A microprocessor has enough energy to control a transistor but not to a run motor directly. Thus, the transistor is used between the microprocessor and the motor. A transistor has three legs called the base, collector, and emitter. The base controls the operation of the transistor. These two transistors discussed below, are really Darlington transistors. For the examples shown here, we will represent them as being typical transistors.

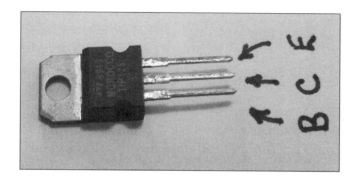

Shown here are two types of transistors: NPN (TIP102) and PNP (TIP125). "B" is the base, "C" is the collector, and "E" is the emitter.

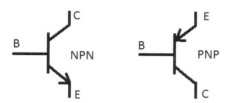

To operate the transistor as a switch, the base is set to 5v or 0v by the microprocessor. When the transistor is on, the collector voltage is about 0.7v higher than the emitter is. The power used by the transistor would be the current through it times this voltage of 0.7v.

Pertaining to the NPN:

If B = 5v, then C and E are connected. The transistor switch is closed.

If B = 0v, then C and E are disconnected. The transistor switch is open.

For the PNP, it is the opposite:

If B = 5v, then C and E are disconnected. The transistor switch is open.

If B = 0v, then C and E are connected. The transistor switch is closed.

The circuit to control a motor would look like this. The AVR would be connected to the resistor, which limits the current going into the transistor. If the transistor is always on, the motor would run at full speed. To control the speed of the DC motor, you can employ pulse-width modulation (PWM).

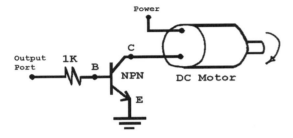

## Relay

A relay is a mechanical switch controlled by a signal voltage. Being mechanical in nature, they can suffer from maintenance issues like corrosion. They may be configured as SPST, SPDT, DPST, or DPDT.

This relay is a DPDT (Double Pole, Double Throw). It is controlled by the inputs at C. If the input is energized by a voltage, then the switches X are connected to either A or B. When not energized, the switch is in the other position. The DPDT has two separate switches, although they move at the same time. In the picture, pin C is located by its greater pin spacing.

 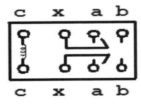

## Photo Resistor

A photo resistor is a resistor that changes its value with light intensity. It is a passive two-legged device. The greater the light, the less resistance it has; with less light, its resistance is greater. A microprocessor like the AVR cannot measure resistance directly, so instead we use a voltage divider circuit to convert resistance into voltage. This voltage is analog and would be sent to an A/D converter or to a comparator such as an LM339. This picture is of a photo resistor.

In the following circuit, the variable resistor R2 is the photo resistor. The output voltage = $5 \times R2/(R2 + R1)$ volts. The value of R1 should be set approximately to the average value of R2 for best sensitivity. The output would be connected to the A/D converter.

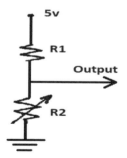

## Potentiometer

A potentiometer is a variable resistor. The potentiometer can be rotated to increase or decrease the resistance between the middle wiper and one of the two outer connectors. These are handy when you do not know exactly what resistance is desired in a circuit or when the resistance needs to be varied.

To get a variable voltage, connect 5v and GND to the two outer pins. Then, the middle pin is some voltage between 5v and 0v corresponding to the angular position of the shaft. The potentiometer converts angular position to a voltage. It is also a great way for a user to make a selection. For example, you could rotate the knob to select a song to play on a CD. For this to be useful, the computer needs to display back to the user the value currently selected as the user turns the potentiometer. This is a picture of a potentiometer and a trim potentiometer.

## Thermal Resistor

A thermal resistor is a resistor that changes its value with temperature. It is a passive two-legged device. As the temperature changes, so does the resistance. A thermal resistor with a smaller thermal mass is able to respond more quickly to temperature changes. Connected to a computer, you could detect fires, temperature extremes, control room temperatures, etc.

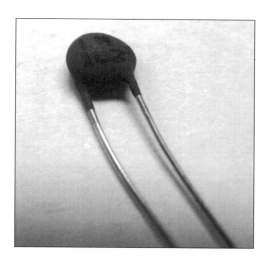

As said before, a microprocessor like the AVRs cannot measure resistance directly, so instead use a voltage divider to convert resistance to voltage. This voltage is analog and would be sent to an A/D converter or to a comparator such as an LM339. For the circuit shown next, R2 represents the thermal resistor. The output going to the A/D equals $5 \times R2/(R2 + R1)$ volts. The value of R1 should be set to the average value of R2 for best sensitivity.

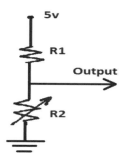

## Temperature Sensor, LM35

This active sensor produces a voltage linearly proportional to the temperature. It needs to be connected to power and ground. The output voltage is typically sent to an A/D converter.

# IR Sensors and Diodes

Infrared sensors can be used to detect the presence of an object or to measure distance to an object. The IR sensor is comprised of an infrared emitting diode that emits infrared light and a matching infrared detecting diode.

### Diodes

Diodes, like resistors, have 2 points of connection but, for the most part, pass current in only one direction—the forward direction. In the forward direction, the resistance is low, and in the reverse direction, the resistance is high; thus, the current is essentially allowed to flow in the forward direction only. The forward direction is denoted by the direction of the triangle. If there is a forward current flowing, then there is a voltage drop of about 0.7 volts across the diode. When connecting the diode, the negative side is the flat side or the banded side.

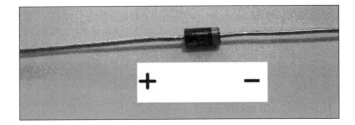

The graph shows the relationship between the diode's current and voltage. The arrow denotes the forward direction of the current.

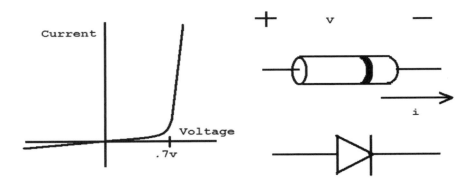

Some diodes emit or detect infrared light as used in a remote control. Diodes can be used to convert alternating current to direct current. A half wave and full wave rectifier converts A.C. to D.C. The half wave rectifier, as shown in this picture, takes an A.C. voltage source and allows the current to flow in one direction thus eliminating the negative voltages. Although the voltage of the D.C. is variable in this circuit, it only flows in one direction (which is the definition of direct current).

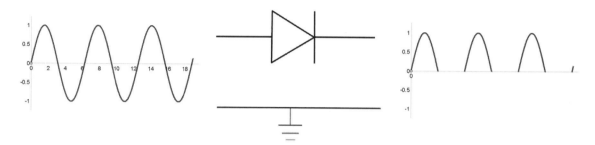

The full wave rectifier, shown next, does not remove the negative voltage but rather negates it to make it a positive voltage.

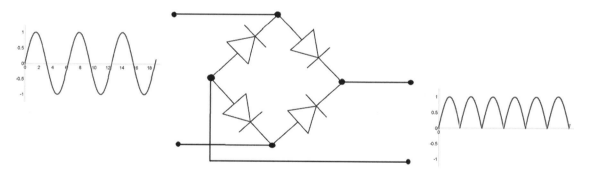

### IR Pairs

Diode IR pairs can be used to emit and detect infrared light. The emitter diode (Jameco part # 106526 or 372817) emits a 940 nm infrared wavelength of light when current flows through it (up to 50 mA). Higher currents are allowed for a very short time but could burn it out. Less current will make for a weaker signal. Do not put a sustained 5v across the emitter, because it will draw too much current and burn out. Limit the current with a resistor. In this picture, the top is the emitter and the bottom is detector.

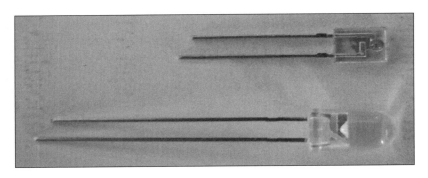

The next circuit shows how to wire the emitter and detector. Using a 5v source, a 50 mA limit, and 0.7v across the emitter yields a current limiting resistor of about 86 ohms. It can be more or less depending on the resistors at hand.

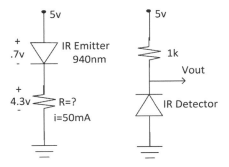

The matching detector diode (Jameco part # 112168) conducts when it sees 940 nm. It has a high resistance when it does not see 940 nm. Vout of the detector circuit is low when the emitter is seen; otherwise, it is high. Emitters and detectors can look very much alike. Some detectors have a very slight black dot in its center. Some emitters have a square shape and are easily distinguished.

### Slotted Optical IR Switch

Optek makes a slotted optical switch which includes the emitter and detector in one 4-pin package (part # OPB660N). Fairchild also makes one (part # H22A1). Both require pull up and current limiting resistors. When an object passes within the slot, the detector does not see the emitter.

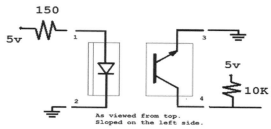

150

5v

1

3

5v

10K

2

4

As viewed from top.
Sloped on the left side.

### IR Distance Sensor

These IR sensors can measure distance to an object within a given range and are very easy to work with. Sharp makes several. One is Sharp's GP2Y0A41. It produces an analog voltage of 3.1v at 4cm down to 0.3v at 30cm. It requires a supply of 5v. The output typically would be connected to one of the A/D channels on the AVR. As for the software, simply set up and read the A/D channel that it is connected to.

These lines of C code show how to set up and read channel 0 on the ATmega328.

```
DIDR0 = 0x3f;                  // Digital input disable
ADCSRA = 0xe0;                 // Continually convert A/D
ADMUX = 0x20;                  // Channel 0, pin 23
delay_ms(1);                   // Wait for conversion
unsigned char temp = ADCH;     // Read result
```

## Ultrasonic Sensor

The HC-SR04 ultrasonic sensor can be used to measure distances from 2cm to 4m. The sensor requires 5v to operate. To start a measurement, set the trigger pin high for 10 usec. Then wait for the echo pin to go high. Measure the time the echo pin is high. This echo measurement is the time it takes sound to go down to and back from the object. From this, the distance can be calculated.

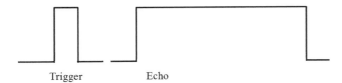

Trigger          Echo

Distance = (speed of sound) × (High Time)/2. Where speed of sound = 343 ms = 1125 ft/s.

This code shows how to find the high time from the HC-SR04 sensor. It loops while the echo line is high. There are 5 cycles per loop and 5 cycles takes 5 usec at a frequency of 5 MHz. Register r17 counts the number of loops. The high time = r17 × 5 usec.

```
// HC-SR04 Sensor Code. Result (high time) will be in r17.
// Portb 0, input for echo
// Portb 1, output for trigger
.org   0x0000
       ldi     r16,0b10
       out     ddrb,r16      ;Set direction of portb
       out     portb,r16     ;Send out trigger pulse for 10 usec.
       nop
       nop
       nop
       nop
       nop
```

```
        nop
        nop
        nop
        nop
        nop
        ldi    r17,0
wfh:                        ; Loop until a high is seen.
        in     r16,pinb   ; Read output of sensor.
        andi   r16,0b01   ; Isolate pin 0
        breq   wfh         ; Wait for high

wfl:                        ; Loop until a low is seen.
        inc    r17         ; [1] Count loops to determine high time.
        in     r16,pinb   ; [1] Read output of sensor
        andi   r16,0b01   ; [1] Isolate pin 0
        brne   wfl         ; [2] Wait for low
// High Time = 5 usec per loop * r17 (loops) at f = 1 MHz
wait:
        rjmp   wait
```

This picture is of the Parallax ultrasonic sensor.

Parallax also makes an ultrasonic sensor very similar to that of the HC-SR04. It works in the very same fashion with the exception that it has 3 pins rather than 4. Here the echo and trigger pin have been combined into the same pin called the signal pin. The AVR needs to connect an I/O pin (like pb0) to the signal pin. To start operation, pb0 needs to be configured as output, then pb0 needs to send out a high pulse for 2 us to 10 us. Next, the AVR must configure pb0 as input and measure the resulting pulse returning on the signal pin. Just like the HC-SR04 ultrasonic sensor, the distance to the object is determined from the same equation.

Distance = (speed of sound) × (High Time)/2. Where speed of sound = 343 ms = 1125 ft/s.

## GPS Sensor

This GPS sensor (part number PMB648) can be used to obtain information as time, latitude, longitude, altitude, speed, and direction. It requires 5v on the red wire, and GND on the black wire. The yellow wire sends out an ASCII data string every second. The string follows the NMEA 0183 standard. It starts with a '$' followed by the string type and data. The serial string is transmitted at 4800 baud. An example of this serial string follows:

$GPGGA,184453.000,4208.1006,N,07554.5916,W,1,05,2.5,247.1,M,-33.9,M,,0000*66

In the string, the first number represents UTC time (18hr, 44min, 53sec). The second item is 42 degrees North latitude followed by 75 degrees West longitude. These pictures show the GPS and its connection to a development board. The connections and software are relatively easy.

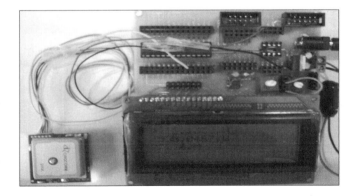

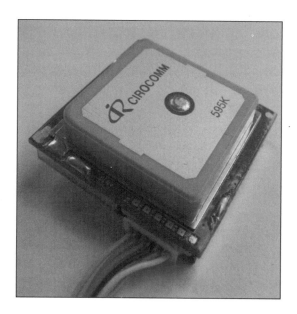

The following main program receives values from the GPS sensor and sends them to the LCD using classes. The include files contain the definition of each class. The line in the program, class serial328 gps, declares gps as a variable of class serial328. The line in the program, gps.rx(), calls the subroutine rx in that class, which is to receive a character from the gps. The files "lcd_16x2spi_cpp.h" and "serial_328cpp.h" can be found in Chapter 2 in the "SPI Software Implementation" section.

```
//---------------------------------------------
#include <avr/io.h>
#include "\avr c programs\lcd_16x2spi_cpp.h"
#include "\avr c programs\serial_328cpp.h"

int main(void)
{   class lcd16x2 lcd;
    class serial328 gps;
    gps.init_rx(25);          // initialize usart to 4800 baud
    lcd.clear();              // clear lcd

    do{
        do {} while ( gps.rx()  != '$' );             // Wait for start of string
        for (int i=0;i<16;i++) lcd.da( gps.rx() );  // Report the first 16 character
      } while (1);
}
//---------------------------------------------
```

## Touch Pad

Touch pads are used to input information into a microprocessor. Often they are placed on top of a picture or LCD screen. The touch pad used here is resistive while others are capacitive. Capacitive touch pads are more expensive but support multi-touch. Our resistive touch pad has 4 wires (y1, x2, y2, x1) emanating from it. It can be thought of as two variable resistors at right angles to each other. One resistor in the x direction, and one resistor in the y direction. By applying 5v at x1 and 0v at x2, you can measure the voltage at y2 (or y1) which will be proportional to the x position.

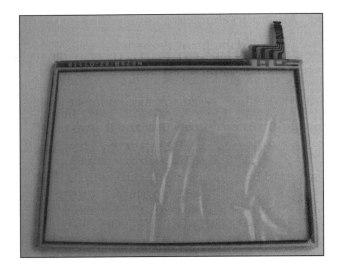

The diagram on the left shows how to read the x position, and the diagram on the right shows how to read the y position.

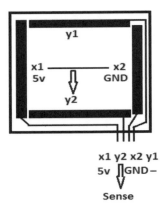

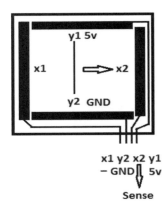

Assuming the A/D is on Port C, then connect the following.

pb6 connects to y1.

pc1 connects to x2 with optional 10k pull down to ground.

pc2 connects to y2 with optional 10k pull down to ground.

pb7 connects to x1.

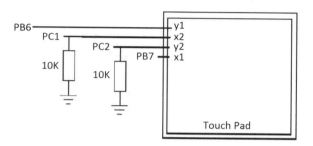

To read the x position, apply 5v to x1 (pertains to left side of screen) and 0v to x2 (pertains to right side of screen). When the pad is pressed, a voltage will be present on y2 (and y1). This voltage is proportional to the x position. The value of the voltage is closer to 5v if your finger is closer to x1 (5v) than x2 (0v). If not pressed, then the voltage read is near zero if using the external pull-down resistor. Without the pull-down resistor, the voltage read will be 5v when not pressed.

Reading both x and y positions at once is a bit tricky, because you must set up the I/O configuration to read x, then switch the I/O configuration to read y. Here the direction of the pins of the AVR are not static but change when reading a coordinate. A sample program is illustrated in Chapter 4, "Projects Using the AVR."

One interesting fact about the touch pad is that when pressed, the resistance (and voltage) does not go immediately to the final value but rather it ramps up from zero to the final value. This means that if you read the value too soon, you will get a value not yet finalized. When the pad is pushed, the signal needs time to settle (on the order of a 1/10 sec). One approach is to read multiple values until the values settle down.

Port b is used to drive the 5v rather than using port c, because port b has better driver capabilities than port c. Use the optional 10K pull-down resistors if you want to have the values read to be near zero when the touch pad is not pressed.

## Accelerometer

The accelerometer mma7361 is made by Freescale. The sensor produces a voltage proportional to the acceleration with an offset. For example, if the acceleration = 0, then the output voltage might be 1.6v. If the acceleration is negative, the voltage would drop from that or increase from that when the acceleration is positive. This device measures the acceleration in the x, y, and z direction. To use it, make g select = Low, sleep = high (off), apply power, and then measure output. See data sheet from Freescale.

| MMA7361 Pin | Meaning |
| --- | --- |
| g select | Maximum g value. 0=1.5g, 1=6g |
| Sleep | Sleep mode. Active low. |
| 0g Detect | Active high when all 3 axes are in free fall. |
| Vss | GND |
| Vdd | Supply Voltage: 2.2-3.6v |
| X Out | X Acceleration |
| Y Out | Y Acceleration |
| Z Out | Z Acceleration |

## Magnetic Field Sensor

The HMC1001 magnetic field sensor produces a voltage proportional to the alignment with the Earth's magnetic field. Its output is on the order of 1 or 2 millivolts, and being so small, it is not practical to send the output directly to an A/D on the AVR. You would need to send its output to an instrumentation operational amplifier. See data sheet from

Honeywell. The pins on this part are closer than 1/10 inch apart so a breakout PCB board was made, as shown in this picture.

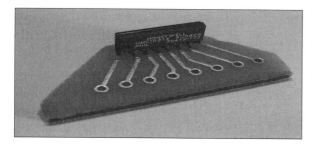

## Comparator

A comparator is not a sensor, but it can be used to compare voltages and then output a result based on which voltage is larger or smaller. This output line is either high or low. Often the input comes from an analog sensor and the digital output goes to a digital input pin on the AVR. The AVR would then do something according to the sensor value. Typical uses include temperature control, speed control, etc.

How does using a comparator differ from that of an A/D converter? They can both read an analog voltage and can make decisions. The A/D converts the analog signal to its digital equivalent whereas the comparator has a single output line that represents the comparison between the input analog signal and a reference voltage.

The ATmega328 has an internal A/D and also an internal comparator. The LM339 is a comparator chip. In this picture, the pin numbers are underlined. The rules that govern the LM339 are:

When IN+ > IN− the output is disconnected. Output = $V_x$

When IN+ < IN− the output is connected. Output = 0

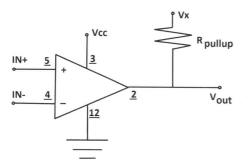

Connected means that the output (or pin 2) is connected to pin 12, which is ground. Disconnected means that pin 2 is not connected to anything inside the chip; therefore, the output takes on the value of $V_x$ which is generally set to 5v. The ability for the LM339 to be disconnected is due to the fact that the comparator has an open collector output. There is no single correct value for $R_{pullup}$ but 1K ohms generally works fine. Because the output is often sent to an input port on the AVR, we will set $V_x$ to 5v. Thus if IN+ is greater than IN− then the output = 5v. On the other hand, if IN+ is less than IN− then the output = 0v. An example of this is shown in the next picture.

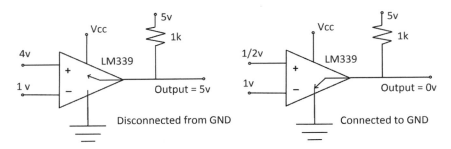

There are 4 separate comparators on one LM339 chip, as shown by the pin out. Power and ground must be supplied. Vcc may be any voltage from 2 to 36v.

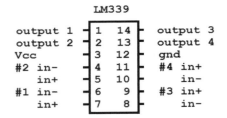

Where do IN+ and IN− come from? In a typical application, IN− would be a fixed reference voltage set by a voltage divider and IN+ would come from a sensor. The next diagram shows how to use a photo resistor to determine when the light in a room is bright or dark. IN− is set to 2.5 volts by the set point potentiometer. The sensor is the photo resistor and R1 is set to be the average value of the sensor. When the light changes, so will the resistance of Rp and thus IN+. The equation for IN+ is: IN+ = 5 × Rp/(Rp + R1) volts.

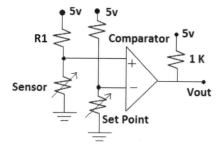

In a dark room, the light decreases on the photo resistor, its resistance increases past its average, and thus IN+ increases past 2.5v (see equation). Remember we made R1 equal to the average of Rp. With IN+ > IN− the output is 5v. In a bright room, the light increases on the photo resistor, then its resistance decreases past its average, and thus IN+ decreases past 2.5v. With IN+ < IN− the output is 0v.

How is the hardware integrated with the software on the AVR? Connect the output of the prior circuit to port b bit 0. Here is the code to wait and read port b bit 0 while there is above average light in the room.

```
bright:
        in    r16,PINC      ; read all 8 bits
        andi  r16,0b01      ; preserve bit 0
        breq  bright        ; branch if 0
```

A comparator can be used for temperature control. Here a temperature sensor (LM35 or thermal resistor) is connected to the input of a comparator which in turn controls a relay which controls the furnace without the use of a microprocessor. It also does not, but could, employ the important aspect of hysteresis. Hysteresis prevents the furnace from cycling on and off too quickly.

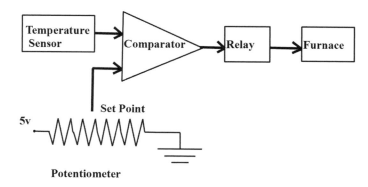

# LCD Displays

LCD screens are used to present information to the user of the device. When designing a system, keep in mind that text screens compared to graphics screens are cheaper, easier to interface, and use less memory. LCDs that use SPI or I2C use fewer I/O connections, which is a significant advantage.

## Text Screens Using Parallel Data

A text LCD such as the HDM20416 LCD display can be connected to a microprocessor. This screen has 4 text lines with 20 characters per line. This text display uses 8 wires for data and 2 wires for control. Thus you need an 8-bit port for the data and another 2 bits for control.

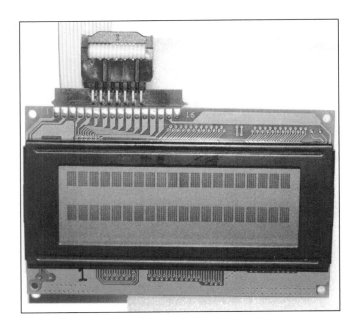

The software to drive the display requires understanding the timing diagrams of the LCD. The simplified timing diagram shows that the RS (data/instruction) line has to be valid first. Making this line low means that an instruction will be on the data lines, whereas making this line high means that data to display will be on the data lines. If RS is high then the ASCII character found on the data lines will be displayed. If RS is low then the data lines contain an instruction such as clear the screen.

As shown by the timing diagram, the R/W line is to be set low when writing instructions or data to the LCD. This line can be wired low if you only write to the LCD. Next, the enable line is brought high, then the data must be sent out onto the data lines. The last step in the process is to drop the enable line. The data will be displayed or the instruction carried out. The signals may now be released. Refer to the following subroutine: void lcd_da (int data) to see how this is done. The following image shows the HDM20416 text screen timing diagram.

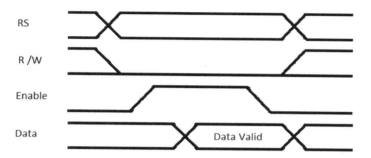

This code example shows how to interface the HDM20416 text LCD to the ATmega328. First connect the hardware as stated here, and then write the code to match the hardware. The #define makes the code easier to maintain and change. The main program needs to call the lcd_init() subroutine first and then the lcd_da() subroutine to send a character to display. In the last two subroutines, the action occurs when the enable line goes from high to low. The falling edge of the enable triggers the LCD to act on the command (many devices are similar).

```
// HDM20416 Example Code for the atmega328. LCD text screen connections.
//          Pin 1,3,5 are gnd.
//          Pin 2 is 5v
//          Pin 4 is RS. 1=data, 0=command. Connected to pc4.
//          Pin 5 is r/w, Always tied low. 1=LCD read, 0=LCD write.
//          Pin 6 is ENABLE LCD connected to pc5.
//          Pin 7-14 is data connected to pd0 to pd7.

// port c bit 5 goes to pin 6 (enable) on lcd.
// port c bit 4 goes to pin 4 (rs) on lcd.
// port d is connected to pins 7-14 (data) on lcd.

#define LCD_ENABLE      0b00100000    // port c bit position for enable
#define LCD_RS          0b00010000    // port c bit position for rs
#define lcd_ctrl_port   PORTC         // control port assignment
#define lcd_ctrl_ddr    DDRC
#define lcd_data_port   PORTD         // data port assignment
#define lcd_data_ddr    DDRD
```

```c
#include "avr/io.h"
#include "delays_1mhz.h"

void lcd_init ();                       // Initializes the screen.
void lcd_si (int instruction);          // Sends the instruction to the LCD
void lcd_da (int data);                 // Displays the ASCII character on LCD
void lcd_home();                        // Sends cursor to the home position
void lcd_clear();                       // Clears the screen

//-----------------------
int main (void)
{ lcd_init();
  lcd_da('A');                          // Example: displays an A
}
//-----------------------
//-----------------------
void lcd_init ()                        // Initializes the screen.
{    lcd_data_ddr = 0xff;               // sets the ddr registers
     lcd_ctrl_ddr |= LCD_ENABLE;
     lcd_ctrl_ddr |= LCD_RS;

     lcd_si(0x0f);                      // command: turn on LCD
     lcd_si(0x3c);                      // command: data is 8 bits
     lcd_si(0x01);                      // command: clear LCD
     delay_ms(10);
     return;
}//-----------------------
//-----------------------
void lcd_si (int instruction)           // Sends the instruction to the LCD
{
     lcd_ctrl_port &= ~LCD_RS;          // treat as an instruction
     lcd_ctrl_port |=  LCD_ENABLE;      // enable high
     lcd_data_port = instruction;       // send instruction
     lcd_ctrl_port &= ~LCD_ENABLE;      // enable low
     delay_ms(1);
     return;
}//-----------------------
//-----------------------
void lcd_da (int data)                  // Displays the ASCII character on LCD
{    lcd_ctrl_port |=  LCD_RS;          // treat as data
     lcd_ctrl_port |=  LCD_ENABLE;      // enable high
     lcd_data_port = data;              // send data for display
```

```
    lcd_ctrl_port &= ~LCD_ENABLE; // enable low
    delay_ms(1);
    return;
}// – – – – – – – – – – – – – – – – – – – – – –
//– – – – – – – – – – – – – – – – – – – – – – –
void lcd_home ()                    // Goes to the home position.
{   lcd_si(0x03);
}// – – – – – – – – – – – – – – – – – – – – – –
//– – – – – – – – – – – – – – – – – – – – – – –
void lcd_clear ()                   // Clears the screen.
{   lcd_si(0x01);
}// – – – – – – – – – – – – – – – – – – – – – –
```

## Graphics Screens Using Parallel Data

The graphics screen illustrated here is the HDM64gs12L4 graphics screen. It has 128 columns by 64 rows of pixels. It uses 8 wires for data and 6 for control. The graphics screen's timing diagram is very similar to that of the text screen. The subroutines needed to talk to the screen have been created and are listed next. Just include them with your main program.

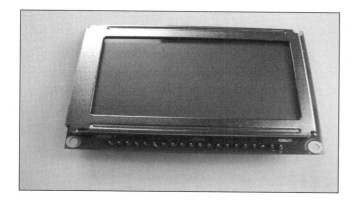

This screen is divided into a left and right half, each half being 64x64 pixels. The signal cs1 selects the left half while cs2 selects the right half. The pixels are written to 8 at a time. It is not possible to write to just one pixel at a time. The screen has 8 pages, each page has 8 rows of pixels. When writing to the screen, select the half (left or right), the column (0 to 63), the page (0 to 7), and then 8 bits (1 byte) are sent representing the 8 vertical pixels at that location, as shown in the picture.

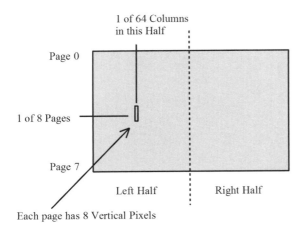

Each page has 8 Vertical Pixels

The software to drive the display requires understanding the timing diagrams from the manufacturer's data sheet. The timing diagram shows that the D/I (Data/Instruction) line has to be valid first. Making this line low means that an instruction will be on the data lines, whereas making this line high means that data will be on the data lines. For example, if line D/I is high, then pixel data will be on the data lines. If line D/I is low, then the data lines contain an instruction such as which column to select. The R/W line is to be set low when writing instructions or data to the LCD. Next, the enable line is brought high, then, if writing, the data must be sent onto the data lines. The last step to complete the process is to drop the enable line. The pixels will be displayed or the instruction carried out. The signals may now be released. Refer to the subroutine void lcd64_sd (int data) to see how this is done. The following image shows the HDM64gs12L4 graphics screen timing diagram.

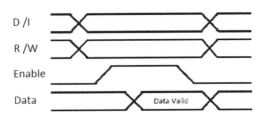

This table and code shows how to interface the HDM64gs12L4 LCD to an ATmega328. First connect the hardware as listed. Port d is for data. Port b is for the control signals. Each control signal is assigned to a pin on port b. The #define makes the code easier to maintain and change.

| HDM64gs12L4 Pin | Signal Name | ATmega328 Pin |
|---|---|---|
| 1 | 5v | pin 7 |
| 2 | GND | pin 8 |
| 3 | Operating Voltage Input, about −4v. | - |
| 4:11 | Data 0:7 | pd 0:7 |
| 12 | CS1 | pb0 |
| 13 | CS2 | pb1 |
| 14 | /Reset | pb2 |
| 15 | R/W | pb3 |
| 16 | D/I | pb4 |
| 17 | Enable | pb5 |
| 18 | Vee, not used | - |
| 19 | back-light, + voltage | - |
| 20 | back-light, − voltage | - |

To use the back-lighting apply 5v in series with a resistor of value 220 to 1K ohms to pin 19. Then connect pin 20 to ground. To get −4v for pin 3, you will need a second voltage source with its positive connected to ground. Another possibility is to use a chip that can produce a negative voltage, such as the TC7660 as shown in this diagram. You would have to rotate the potentiometer to obtain the correct LCD contrast.

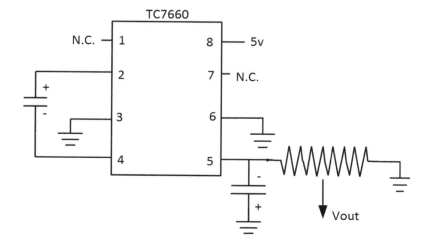

The main program must call subroutine lcd64_init() first. Then it lights a pixel, draws a circle, and draws a line. When writing a pixel, the code first reads the 8 pixels in that page and then alters the 1 pixel in question and finally sends the 8 pixels back out to the LCD. The main program includes the file lcd_64gs12L4_driver328.h which in turns includes the file lcd_64gs12L4_driver.h.

```c
//------------------------------------
//   atmega328_graphics2.c
#include <avr/io.h>
#include "lcd_64gs12L4_driver328.h"

int main(void)
{  lcd64_init();              // Required
   lcd64_clr();               // Clear screen
   lcd64_xy(4,9,1);           // Light 1 pixel
   lcd64_circle(64,32,20,1);  // Draw a circle.

   for (int i=0;i<5;i++)      // Draw lines at random.
     { int x2 = rand()%128;
       int y2 = rand()%64;
       int x1 = rand()%128;
       int y1 = rand()%64;
       lcd64_line(x2,y2,x1,y1,1);
     }
   return 0;
}//---------------------------------
```

Shown next is the first include file called lcd_64gs12L4_driver328.h

```c
//------------------------------------
// lcd_64gs12L4_driver328.h    Driver for the hdm64gs12-L4 screen.
// Works with the atmega328.
// Connect graphics screen. Use potentiometer to set contrast.
// Call lcd64_init() first and then any public subroutines.

// Public Subroutines
void lcd64_init ();         // This must be called first before using lcd.
void lcd64_clr ();
void lcd64_xy (int x, int y, int color);
void lcd64_line (int x1, int y1, int x2, int y2, int color);
void lcd64_circle (int xc, int yc, int r, int color);
unsigned char lcd64_page_read (int c, int p);       // Read 8 pixels.
void lcd64_page_write (int c, int p, int pattern);   // Write 8 pixels.
```

```
// Private Code
void lcd64_col (int col);
void lcd64_page (int page);
void lcd64_sd (int data);
int  lcd64_rd ();

void lcd64_si (unsigned char instr);
void lcd64_lh ();
void lcd64_rh ();
void lcd64_bh ();
void delay_ms (int t);

#define lcd64_ctrl_port          PORTB
#define lcd64_ctrl_port_ddr      DDRB
#define lcd64_data_port_out      PORTD
#define lcd64_data_port_ddr      DDRD
#define lcd64_data_port_in       PIND

#define LCD64_CS1      0b00000001
#define LCD64_CS2      0b00000010
#define LCD64_RST      0b00000100
#define LCD64_RW       0b00001000
#define LCD64_DI       0b00010000
#define LCD64_ENABLE   0b00100000

#define LCD64_COL0     0x40
#define LCD64_PAGE0    0xb8
#define LCD64_ON       0x3f

//------------------------------------
void lcd64_init ()
{   lcd64_ctrl_port_ddr = 0x3f;
    lcd64_data_port_ddr = 0xff;
    lcd64_ctrl_port |=  LCD64_RST;
    lcd64_bh();
    lcd64_si(LCD64_ON);
    lcd64_clr();
}
//------------------------------------
#include "lcd_64gs12L4_driver.h"
//------------------------------------
```

Next is the second and last include file called lcd_64gs12L4_driver.h

```
//------------------------------------
// lcd_64gs12L4_driver.h
// Driver for graphics screen with any AVR.
```

```
// LCD pin out
//pin    1    5v
//        2    gnd
//        3    vo, operating voltage input, about -4v.
//     4:11    data 0:7
//       12    cs1
//       13    cs2
//       14    /reset
//       15    r/w
//       16    d/i
//       17    enable
//       18    vee, not used
//       19    backlight +
//       20    backlight -

//--------------------------------
// Circle at (xc,yc) with a radius of r and color.
void lcd64_circle (int xc, int yc, int r, int color)
{
    int r2 = r*r;
    for (int dx=0;dx<=r;dx++)
    {   int dx2 = dx*dx;
        int dy = sqrt(r2-dx2);           // delta y
        lcd64_xy(xc+dx,yc+dy,color);     // above and below yc
        lcd64_xy(xc+dx,yc-dy,color);
        lcd64_xy(xc-dx,yc+dy,color);     // above and below yc
        lcd64_xy(xc-dx,yc-dy,color);
    }

    for (int dy=0;dy<=r;dy++)
    {   int dy2 = dy*dy;
        int dx = sqrt(r2-dy2);           // delta y
        lcd64_xy(xc+dx,yc+dy,color);     // above and below yc
        lcd64_xy(xc+dx,yc-dy,color);
        lcd64_xy(xc-dx,yc+dy,color);     // above and below yc
        lcd64_xy(xc-dx,yc-dy,color);
    }
}//--------------------------------
//--------------------------------
// Line from (x1,y1) to (x2,y2) with color.
void lcd64_line (int x1, int y1, int x2, int y2, int color)
{   int xdelta = x2-x1;
    int ydelta = y2-y1;
```

```
      int x,y;
      int xlow,xhigh,ylow,yhigh;

      float m = (float)(ydelta)/(xdelta);

      if (xdelta<0) xdelta = -xdelta;
      if (ydelta<0) ydelta = -ydelta;

      if (ydelta>xdelta)
         { if (y2>y1)    {yhigh=y2;ylow=y1;} else {yhigh=y1;ylow=y2;}
            for(int y=ylow;y<=yhigh;y++)
              { x = x2-(y2-y)/m;
                lcd64_xy(x,y,color);
              }
         }
      else { if (x2>x1)    {xhigh=x2;xlow=x1;} else {xhigh=x1;xlow=x2;}
            for(int x=xlow;x<=xhigh;x++)
              {     y = y2+(x-x2)*m;
                    lcd64_xy(x,y,color);
              }
         }
}//------------------------------------
// ------------------------------------
// Pixel at (x,y) is set to the color.
void lcd64_xy (int x, int y, int color)
{    int page=y/8, remainder=y%8, pattern;

     lcd64_col(x);
     lcd64_page(page);
     pattern = lcd64_rd();      //Read old dots. Device requires double read!
     pattern = lcd64_rd();      //Read old dots. Device requires double read!

     lcd64_col(x);
     lcd64_page(page);
     if (color==1)    pattern |=  (1<<remainder);      //add 1 more dot, ON
            else    pattern &= ~(1<<remainder);      //add 1 more dot, OFF
     lcd64_sd(pattern);
}//------------------------------------
// ------------------------------------
// Return the 8 bits (pixels) at (column,page).
unsigned char lcd64_page_read (int c, int p)
{  lcd64_col(c);
   lcd64_page(p);
   lcd64_rd();                //requires a double read
   return (lcd64_rd());
}//------------------------------------
```

```
//---------------------------------
// Writes the 8 bits (pixels) at (column,page).
void lcd64_page_write (int c, int p, int pattern)
{   lcd64_col(c);
    lcd64_page(p);
    lcd64_sd(pattern);
    return;
}//---------------------------------
//---------------------------------
void lcd64_sd (int data)    // Send data to screen.
{   lcd64_ctrl_port |=   LCD64_DI;
    lcd64_ctrl_port &= ~LCD64_RW;
    lcd64_ctrl_port |=   LCD64_ENABLE;
    lcd64_data_port_out = data;
    lcd64_ctrl_port &= ~LCD64_ENABLE;
}//---------------------------------
//---------------------------------
// Returns the 8 bits at the current (col,page) on the lcd64
int lcd64_rd ()
{   lcd64_data_port_ddr = 0x00;
    lcd64_ctrl_port |=   LCD64_DI;
    lcd64_ctrl_port |=   LCD64_RW;
    lcd64_ctrl_port |=   LCD64_ENABLE;
    int data = lcd64_data_port_in;         //lcd_data_port;
    lcd64_ctrl_port &= ~LCD64_ENABLE;
    lcd64_data_port_ddr = 0xff;
    return (data);
}//---------------------------------
//---------------------------------
// Clears the screen of the lcd64.
void lcd64_clr ()
{    for (int col=0;col<128;col++)
        { lcd64_col(col);
          for (int page=0;page<8;page++)
             { lcd64_page(page);
               lcd64_sd(0);
             }
        }
}//---------------------------------
```

```
//---------------------------------------
void lcd64_col (int col)    // Selects column
{   if (col<64) lcd64_lh();
            else {lcd64_rh(); col -= 64;}
    unsigned char instr = LCD64_COL0;
    instr |= col;
    lcd64_si(instr);
}//---------------------------------------
//---------------------------------------
void lcd64_page (int page)    // Selects page
{   unsigned char instr = LCD64_PAGE0;
    instr |= page;
    lcd64_si(instr);
}//---------------------------------------
//---------------------------------------
void lcd64_lh ()        // Select left half
{   lcd64_ctrl_port |=  LCD64_CS1;
    lcd64_ctrl_port &= ~LCD64_CS2;
}//---------------------------------------
//---------------------------------------
void lcd64_rh ()        // Select right half
{   lcd64_ctrl_port |=  LCD64_CS2;
    lcd64_ctrl_port &= ~LCD64_CS1;
}//---------------------------------------
//---------------------------------------
void lcd64_bh ()        // Select both halves
{   lcd64_ctrl_port |=  LCD64_CS1;
    lcd64_ctrl_port |=  LCD64_CS2;
}//---------------------------------------
//---------------------------------------
void lcd64_si (unsigned char instr)   // send instruction
{   lcd64_ctrl_port &= ~LCD64_DI;
    lcd64_ctrl_port &= ~LCD64_RW;
    lcd64_ctrl_port |=  LCD64_ENABLE;
    lcd64_data_port_out = instr;
    lcd64_ctrl_port &= ~LCD64_ENABLE;
}//---------------------------------------
//---------------------------------------
void delay_ms (int t)               // t = # msec.
{ for (int j=0;j<t;j++)
    for (unsigned char i=0;i<54;i++);   // 1ms at 1MHz.
  return;
}//---------------------------------------
```

## Text and Graphics Screens Using SPI

LCD screens that use SPI communications generally require fewer I/O lines from the microprocessor. This is a huge savings in terms of board space, cost, and the ability to incorporate multiple devices. Newhaven makes a text SPI LCD (NHD-0420D3Z) and a graphics SPI LCD (NHD-C12832A1Z) as shown in the following picture.

Both displays operate on 3.3v. A complete example program using the text screen is listed in Chapter 2 in the "SPI Software Implementation" section. This C++ example uses classes to talk to the SPI graphics screen. The #defines and this table show how to connect the AVR to the LCD.

| AVR | SPI Graphics LCD |
| --- | --- |
| port d bit 4 | CS |
| port d bit 2 | DI |
| port d bit 1 | SC |
| port d bit 0 | MO |
| Reset | Reset |
| 3.3v | Power |
| GND | GND |

```
//----------------------------------------
// lcd_128x32spi_cpp.h
class lcd128x32
{ // LCD uses 3.3v !!!
   // This can use any port. It does a software implementation of SPI.
   // RESET of LCD is tied to reset of AVR

    // Port used by LCD
    #define PORT    PORTD
    #define DDR     DDRD

    // Bits in port for control lines
    #define CS                0b0010000      // chip select
    #define DI                0b0000100      // data,/instruction
    #define SC                0b0000010      // spi clock
    #define MO                0b0000001      // mosi, disallows use of usart rx
// #define LCD16x2SPIS_MO    0b10000000     // mosi, if used frees up usart rx

   private:
       void col (unsigned char x);
       void dump ();
       void setpage (unsigned char p);
       void dat (unsigned char data);
       void cmd (unsigned char data);
       void out (unsigned char data);
       unsigned char lcd[4][128];
   public:
       lcd128x32();
       void xy (unsigned char x, unsigned char y, unsigned char color);
       void line (int x1, int y1, int x2, int y2, int color);
       void clr ();
};
//----------------------------------------
//----------------------------------------
// Sets page p. 0,1,2, or 3. A page is a row of 8 pixels.
void lcd128x32::setpage (unsigned char p)
{    cmd(0xB0+p);
}
//----------------------------------------
//----------------------------------------
// Displays the pixel at x,y in the range of 128x32.
void lcd128x32::xy (unsigned char x, unsigned char y, unsigned char color)
{    unsigned char p,r,rvalue;
```

```
        // Find page and row.
                if (y>=8*3) { p = 3; r = y - 8*3; }
        else    if (y>=8*2) { p = 2; r = y - 8*2; }
        else    if (y>=8*1) { p = 1; r = y - 8*1; }
        else    if (y>=8*0) { p = 0; r = y - 8*0; }

        // r = row number within page. Find pixel to set.
        if (r==7) rvalue = 0x80;
        if (r==6) rvalue = 0x40;
        if (r==5) rvalue = 0x20;
        if (r==4) rvalue = 0x10;
        if (r==3) rvalue = 0x08;
        if (r==2) rvalue = 0x04;
        if (r==1) rvalue = 0x02;
        if (r==0) rvalue = 0x01;

        // Set or reset pixel.
        if (color==1) lcd[p][x] |= rvalue;
        if (color==0) lcd[p][x] &=~rvalue;

        // Locate column.
        unsigned char col1, col0;
        col1 = x & 0xf0;        // keep upper 4 bits
        col1 /= 16;             // roll right 4 bits
        col1 |= 0x10;           // set bit 4.
        col0 = x & 0x0f;

        // Set page, column, change pixel.
        cmd(0xB0+p);                // set page
        cmd(col1);                  // set col. upper 4 bits
        cmd(col0);                  // set col. lower 4 bits
        dat(lcd[p][x]);             // send data
}
//-----------------------------------------------
//-----------------------------------------------
void lcd128x32::col (unsigned char x)
{   // Locate column.
    unsigned char col1, col0;
    col1 = x & 0xf0;        // keep upper 4 bits
    col1 /= 16;             // roll right 4 bits
    col1 |= 0x10;           // set bit 4.
    col0 = x & 0x0f;
```

```
    // Set column.
    cmd(col1);                      // set col. upper 4 bits
    cmd(col0);                      // set col. lower 4 bits
}
//------------------------------------------
//------------------------------------------
// Line from (x1,y1) to (x2,y2) with color.
void lcd128x32::line (int x1, int y1, int x2, int y2, int color)
{   int xdelta = x2-x1;
    int ydelta = y2-y1;

    int x,y;
    int xlow,xhigh,ylow,yhigh;

    float m = (float)(ydelta)/(xdelta);

    if (xdelta<0) xdelta = -xdelta;
    if (ydelta<0) ydelta = -ydelta;

    if (ydelta>xdelta)
        { if (y2>y1) {yhigh=y2;ylow=y1;} else {yhigh=y1;ylow=y2;}
          for(int y=ylow;y<=yhigh;y++)
            {   x = x2-(y2-y)/m;
                xy(x,y,color);
            }
        }
    else { if (x2>x1) {xhigh=x2;xlow=x1;} else {xhigh=x1;xlow=x2;}
          for(int x=xlow;x<=xhigh;x++)
            { y = y2+(x-x2)*m;
              xy(x,y,color);
            }
        }
}//--------------------------------------
//------------------------------------------
// clear lcd memory and screen.
void lcd128x32::clr ()
{    for (int p=0;p<4;p++)
        for (int c=0;c<128;c++)
            lcd[p][c] = 0;
    dump();
}
//------------------------------------------
//------------------------------------------
// dump lcd memory to lcd
```

```cpp
void lcd128x32::dump ()
{   for (int p=0;p<4;p++)
        {   cmd(0xB0+p);                // set page
            cmd(0x10);                  // set col 0
            cmd(0x00);
            for (int c=0;c<128;c++)
                dat(lcd[p][c]);
        }
}
//-----------------------------------------
//-----------------------------------------
// Initialization For LCD controller
lcd128x32::lcd128x32()
{
    DDR   |= CS;
    DDR   |= DI;
    DDR   |= MO;
    DDR   |= SC;

    cmd(0xA0);    // sets display RAM addr.
    cmd(0xAF);    // display on

    cmd(0xC0);    // select COM output scan direction
    cmd(0xA2);    // sets LCD drive voltage bias
    cmd(0x2F);    // internal power supply mode
    cmd(0x21);    // resistor ratio           //
//  cmd(0x81);    // Vo output voltage         // makes a little darker background
//  cmd(0x3F);    // Vo output voltage         // makes a very dark background
    clr();
}
//-----------------------------------------
//-----------------------------------------
// Sends data to LCD. A0=0 for instruction, A0=1 for data
void lcd128x32::dat (unsigned char data)
{   PORT |=  DI;  //  A0 = 1;
    out(data);
}
//-----------------------------------------
//-----------------------------------------
// Sends a command to LCD. A0=0 for instruction, A0=1 for data
void lcd128x32::cmd (unsigned char data)
{   PORT &= ~DI;  // A0 = 0;
    out(data);
}
//-----------------------------------------
```

```
//----------------------------------------
// Uses a port and manually sends out the bits using SPI
// Rising edge of clk. MSB first.
void lcd128x32::out(unsigned char data)
{   PORT &= ~CS;                // CS = 0;
    for(int i=0; i<8; i++)
    {   PORT &= ~SC;            //SCL = 0;
        if (data&0x80)
                PORT |=  MO;
          else  PORT &= ~MO;
        data<<=1;
        PORT |= SC;             //SCL = 1;
        //delay if needed
        PORT &= ~SC;            //SCL = 0;
    }
    PORT |=  CS;                //CS = 1;
}
//----------------------------------------
```

## XBee Wireless

The XBee wireless modules allow serial data to be sent wirelessly between two AVRs. The Series 1 modules need no configuration to start up and it transmits up to 100 ft. Series 2 needs more setup but can transmit up to a mile. These modules use 3.3v for Vcc, data in, and data out.

| XBee Pin | Purpose |
|---|---|
| 1 | Vcc, 3.3v |
| 2 | UART data out to AVR at 3.3v |
| 3 | UART data in from AVR at 3.3v |
| 10 | GND |

The picture that follows shows the minimum connections needed for the XBee. The bottom XBee receives (on pin 3) a serial data string at 9600 baud from the '328 via a potentiometer. The purpose of this potentiometer is to convert a serial data stream with

5v highs to 3.3v highs. The potentiometer must first be calibrated to produce 3.3v on the middle wiper when 5v is presented on one of the outside pins and ground on the other outside pin. A program is written to send serial data out from pin 3 (port d1) of the '328. The '328 can operate on 3.3v, which would be compatible with the XBee, but the LCD used in the example needed 5v signals to operate. Some LCDs can operate on 3.3v. Be sure to have common grounds and that the XBee does not get any voltage above 3.3v on any pin! Again, note that the AVR can operate on 3.3v and thus no voltage level conversion would be needed. In addition, if the AVR is operating on 5v, it can accept 3.3v as a high input signal.

In the next picture, the ATmega328 is operating at 5v. It is sending out a serial data stream which is reduced to 3.3v by the potentiometer. From there it goes to the XBee sender (bottom XBee) which transmits it wirelessly to the XBee receiver (top XBee). In order to see the data coming through the XBee receiver, it could send the data (out from pin 2) to a terminal window on a PC or to another '328 connected to an LCD. The second '328 could also log the received data.

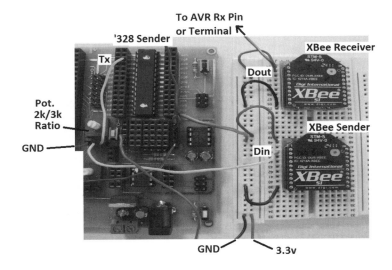

The XBee can only run on 3.3v whereas the '328 can run on 3.3v or 5v, but it must run on 5v if it is connected to a 5v device. It is convenient to run on just one voltage level. This next diagram shows a sensor going into the '328 (on the right). The '328 on the left picks up the data wirelessly and displays it on an LCD. The serial port of both AVRs must be set to 9600 baud to match that of the XBee.

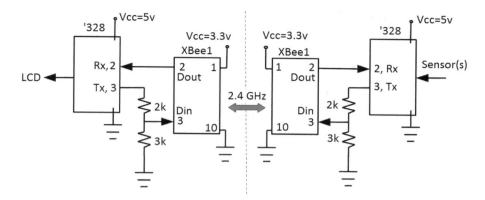

A wire loop back can be used for testing the XBee, as shown in the next picture. Have the '328 on the left side transmit a 'hello'. To see that the 'hello' is received back on the left side, you can have the AVR display it on an LCD text screen or in a terminal window on the PC.

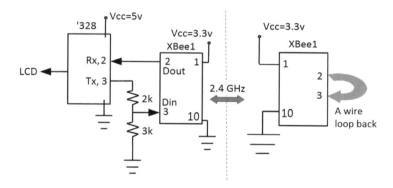

To send out a 'hello' or characters, the main program uses the subroutines found in serial_328.c.

```c
// Code for sending out serial data for the XBee.
#include "serial_328.c"
int main(void)
{   serial_init9600();
    for (int i='a';i<'z';i++) serial_t(i);
}
```

# MOTORS

Motors are useful for moving things and creating motion. This section will look at three types of motors. DC motors spin at a rate determined by the voltage supplied. Steppers move in steps, so they are good for positioning. Servos are good for positioning within a limited angular range. They are easier to use than steppers and are lighter.

## DC Motors

The shaft of a DC motor will rotate with a speed that is proportional to the voltage supplied to the motor. It has 2 wires going to it. To go forward, force one wire to a positive voltage and the other to ground. To go in reverse, switch the voltages. A microprocessor cannot supply enough power to run a motor of any sort, but it can control a switch that in turn connects the motor to power. The next few pictures show the outside and insides of a DC motor.

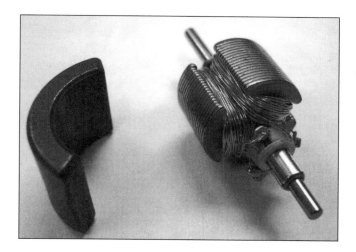

The next diagram shows how to connect a motor to a transistor switch which is controlled by a microprocessor. For an NPN, if B = 5v then the transistor connects C and E and the motor runs. If B = 0v then the transistor disconnects C and E and the motor does not run. In this fashion the transistor acts like a switch.

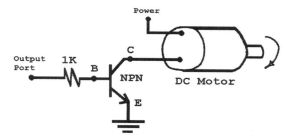

For the microprocessor to control the speed of the motor, it can utilize Pulse Width Modulation or PWM. With PWM the motor is turned on and off very rapidly. The speed is proportional to the on time divided by the on time plus off time. This ratio is also known as the duty cycle. A duty cycle of 75% is on for 3/4 of the period. The motor will speed up during the on time and slow down during the off time, but this is not really noticed since the switching time is very short.

The next circuit shows how to control the speed and direction of a DC motor. It uses two NPN transistors. One transistor for controlling direction and one transistor for PWM or On/Off. Also needed is a relay large enough to handle the currents. When switching direction of the relay, it is highly recommended that you turn off the motor first so that the relay is not burned out when switched.

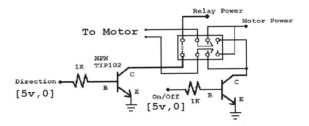

## Full-Bridge or H-Bridge Motor Driver

The Full-bridge, such as the LMD18245, can be used to control a DC or stepper motor. It contains power switches in an H-bridge configuration. Being a single component, it can save you a bit of work compared to assembling numerous components.

## Stepper Motors

A stepper motor is a motor that rotates a shaft in discrete steps. Each step of the motor is a fixed number of degrees. The purpose is to rotate the shaft to a known position rather than spinning the shaft at a known speed. The stepper is used in applications where you want to position something such as a printer head, CD changer, robot arm, etc. This type of motor is usually more expensive and can use a lot of current. It has a limited speed at which it can move. The circuitry to run this type motor is more complex than other motors, and it often requires a microprocessor to drive it. It is well suited for positioning something.

A stepper motor is useful when you need to know the position of an object as it is being moved. The motion may be linear or angular. For example, some printers use a linear stepper motor to move the print cartridge. Stepper motors are also used in machines that assemble printed circuit boards. The stepper rotates in discrete steps. For example a 48 step/revolution motor will move in steps of 360/48 or 7.5 degrees per step. By moving in discrete steps and knowing how many steps have been taken, you then know the current position. When not in motion, the motor holds the current position which also means that it requires energy to do this. Compare the stepper to a DC motor. As a DC motor spins, you typically do not know the angular position of the motor shaft although you may know the angular velocity. Notice the 4 wires coming from the stepper motor and the magnetic poles in the picture of the stepper.

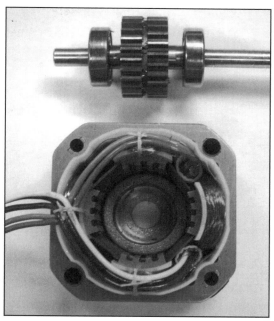

Inside the stepper motor the rotor (the part that rotates) is made up of a permanent magnet. The stator (the part surrounding the rotor) has a number of phases made up of electromagnets. The phases are energized in a given order to make the motor rotate CW or CCW. In this next picture, the stepper motor goes through steps A through C as the motor is sequenced by the windings. Many steppers have a maximum step rate of 50 to 100 steps per second.

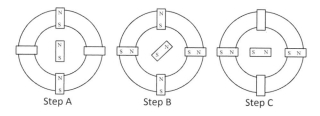

The microcontroller can only output small amounts of current and thus does not have enough energy to run a motor but does have enough power to drive a transistor which in turn can drive a stepper motor.

### How Switching Transistors Work

NPN (a TIP102) and PNP (a TIP125) transistors have 3 legs or connections. The legs are named base, collector, and emitter as shown in the pictures.

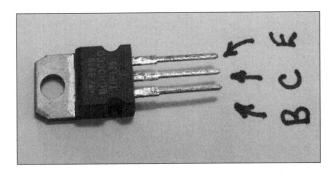
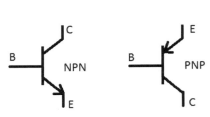

Like a switch, the transistor can be either on (connected) or off (disconnected). This on/off will control the current in the windings of the motor.

For the NPN, if B = 5v then C and E are connected. If B = 0v then C and E are disconnected.

For the PNP, if B = 5v then C and E are disconnected. If B = 0v then C and E are connected.

### Specifics of Driving a Stepper Motor

The stepper motor like a DC motor has a rotor, which rotates, and a stator which is stationary. But unlike the DC motor, the rotor does not have a commutator to bring electricity to the rotor. Thus, the stepper requires less maintenance. To move the rotor, the stator windings change polarity. The direction of rotation is controlled by the order in which the windings are energized. There are generally two types of steppers—bipolar and unipolar. To determine which type you may have, count the number of wires and also test with an inductance meter.

Bipolar motors have only one winding per stator pole. These motors will have 4 lead wires. A push-pull transistor circuit can be used to energize the windings. On the other hand, unipolar motors have two identical sets of windings on each stator pole. These motors may have 5, 6, or 8 wires. The common wires may be connected or completely separate. A simple on/off power transistor will work to energize each winding, because current travels in only one direction in these windings.

### Unipolar Driver Circuit

To wire a unipolar stepper motor, first find the common wires coming out of the motor. There may be 1, 2, or 4 of these. They are to be connected to the positive of the power supply. Use an ohmmeter or inductance meter to locate the common wires. The resistance from A1 to common1 will be half that of A1 to B1. The other windings A1, A2, B1, and B2 will be connected to the collector of an NPN (a TIP102). You will need 4 of these NPNs. The emitter goes to ground. The base is connected to an AVR output pin through a 1k resistor. When the transistor is on, the current runs from the power, through the motor, and then to ground. Thus energizing the winding.

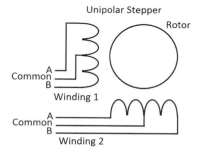

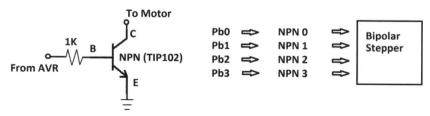

### Bipolar Driver Circuit

The current of the bipolar stepper motor needs to go in both directions. We can use the following push-pull transistor circuit for each wire going to the bipolar motor. When the AVR outputs 5v (on a single bit from the port), the PNP is off and the NPN is on; thus Vcc sources current to one of the motor windings. When the AVR outputs a low, then the PNP is on and the NPN is off; thus GND sinks current from one of the motor windings. The base resistor limits the current going to the base. The voltage going to the motor is limited by the voltage at the base which would be 5v − 0.7v = 4.3v rather than the supply voltage of 5v. If the supply voltage were to increase above 5v this would not translate into a higher voltage to the motor. Using an H-Bridge motor driver (LMD18245) would be a simple way to accomplish higher voltages going to the motor.

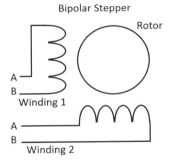

Bipolar Stepper

Rotor

A
B
Winding 1

A
B
Winding 2

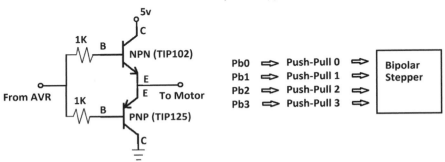

Push-Pull Transistor Circuit for Bipolar Stepper

| Pb0 ⇨ Push-Pull 0 ⇨ | Bipolar |
| Pb1 ⇨ Push-Pull 1 ⇨ | Stepper |
| Pb2 ⇨ Push-Pull 2 ⇨ | |
| Pb3 ⇨ Push-Pull 3 ⇨ | |

### Software to Drive Stepper Motor Circuit

The wires coming out of the stepper may not be in sequential order, so you may need to use an ohmmeter to determine the winding order. To get the motor to rotate, send out on one of the ports the following rotating bit pattern to the transistor circuits of the motor.

```
send bits 1100, then delay (about 1/20 to 1/5 sec)
          0110,      delay
          0011,      delay
          1001,      delay
```

Repeat the sequence to rotate further, or reverse the pattern to switch directions. The delay determines the speed. The stepper has a maximum speed of about 50 to 100 steps per second. The torque decreases as the speed increases. For this stepper code, port b is connected to the transistor circuit.

```
        ldi    r16,0x0f
        out    ddrb,r16    ;data direction register b,
                           ;lower 4 bits are output.

        ldi    r16,0b0011
        ldi    r17,0b0110
```

```
        ldi     r18,0b1100
        ldi     r19,0b1001
top:
        out     portb,r16
        rcall   delay
        out     portb,r17
        rcall   delay
        out     portb,r18
        rcall   delay
        out     portb,r19
        rcall   delay

        rjmp    top
```

Full stepping of the motor involves 2 ones and 2 zeros and walking these bits around in a circle. Half stepping alternates between one and two windings on at a time while rotating. This offers twice the resolution of full stepping. Micro stepping involves using PWM to energize the windings anywhere from 0 to 100% duty cycle. The latter gives the greatest resolution, but it is not shown here.

## Servo Motors

The servo is similar to the stepper in that both motors control the position of the shaft. The stepper shaft can continually rotate whereas the servo generally has a limited range of motion. Often the range of motion is +/− 90 degrees. This periodic signal sent to the servo determines what position the shaft will move to.

Servo waveform:

The signal going to the servo is to be a square wave with a period of 20 msec. The high time of the square wave will range from 1.25 to 1.75 msec., and it is this high time that determines the position.

| High Time | Angle of Servo |
|-----------|----------------|
| 1.25 msec | −90 |
| 1.50 | 0 |
| 1.75 | +90 |

For a servo that has continuous rotation, the high time determines the speed and direction of rotation. It can also be made to stop.

| High Time | Continuous Rotation Servo |
|-----------|---------------------------|
| 1.25 msec | Counterclockwise Fast |
| 1.50 | Stop |
| 1.75 | Clockwise Fast |

The servo's control signal can be generated directly from an output port of the microprocessor or by an LM555 timer as presented in the first chapter. The LM555 resistors that determine the high time could be implemented by a simple potentiometer or even a programmable resistor (DS1804) controlled by a microprocessor, as shown in the next picture. The output of a 555, in the a-stable mode, has a minimum duty cycle of 50%. For that reason, the diagram uses an inverter after the 555. For many servos, the black wire goes to ground, the red to power (5v), and the yellow is the control signal input. It is quite possible to connect an output pin of the microcontroller directly to the control wire of the servo, but the servo should have a separate power supply.

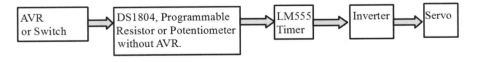

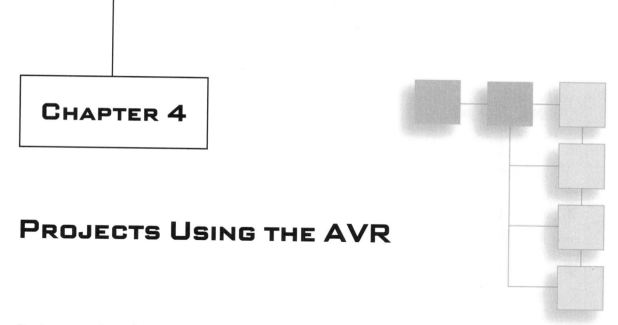

# CHAPTER 4

# PROJECTS USING THE AVR

Projects can be a fun way to gain understanding of the AVR and associated hardware. You can build these projects as shown or perhaps use them as a starting point for something more interesting. Some of the projects here are completely presented while others are presented in various stages of completion.

When working on a project, try to think through all parts of the project to see how feasible the idea is. If one part is particularly challenging or new, you may want to investigate that first to see if it can be done. Complete one step at a time and verify that it works before going on to the next step. Build on prior success. Do not do a number of steps and then try to test it, because it will be more difficult to know which step or steps are failing. Consider using a 3D printer to make mechanical parts. In addition, you may want to design and order a manufactured PCB board for the electronics. Projects introduced in this chapter are as follows.

Frequency Counter

Coin Sorter

Guitar Hero

Morse Code Transmitter

Speed Measurement Using the Timer

GPS Reporting Latitude, Longitude, and Time

Measuring an Incline

Rotating LED Message

Sketch Pad Using a Graphics LCD and Potentiometer

Sketch Pad Using a Graphics LCD and Touch Pad

Tic-Tac-Toe Using a Graphics LCD

Controlling a Servo Motor

The Useless Machine Using a Servo

Sun Locator Using a Servo

Servo with Timer Interrupt Example

Mass Balance Using a Stepper

Combination Lock Opener Using a Stepper

Sleep Mode with Interrupt

## Include Files

Some projects in this chapter use the same subroutines, such as delays and A/D. Instead of listing them multiple times in each project, their location is listed here.

The include file `delays_1mhz.h` is shown here.

The include file `atmega328_ad.h` is shown here.

The include file `touchpad_driver.h` is shown here.

The include file `timer_32.h` is listed in Chapter 2 under "Timer Counter."

The include file `serial_328cpp.h` is listed in Chapter 2 under "USART0 Using the ATmega328 in C."

The include file `lcd_16x2spi_cpp.h` for the SPI Text screen, NHD-C0216CZ, is listed in Chapter 2 under "SPI Software Implementation."

For an example of the text screen, HDM20416, refer to Chapter 3.

For an example of the graphics screen, HDM64gs12L4, refer to Chapter 3.

For an example of the SPI graphics screen, NHD-C12832A1Z, refer to Chapter 3.

## Delay Routines

This subroutine can be used to get a delay of 1 or more milliseconds. It passes an integer, t, representing the number of milliseconds to delay for. It is calibrated using the simulator for a 1 MHz clock. If the AVR is running at 1.2 MHz, change the 54 to a 64. One caveat if you are using AVR Studio 6, be sure to turn off optimization; otherwise, the delay is removed since it is seen as doing nothing. To turn off optimization, click on Project > Properties > Compiler > Optimization and then select None for Optimization Level.

```
// delays_1mhz.h
//------------------
void delay_ms(int t)
{for(int j=0;j<t;j++)
    for (int i=0;i<54;i++);          // 1msec
}//------------------
```

## A/D Routines

These three subroutines can be used to read the A/D on the '328. You need to call the initialization subroutine, ad_ch_init(), first. The subroutine ad_ch() changes the channel and returns a value. The subroutine ad() returns a value using the current channel. Do not forget to connect Vref of the A/D. To use the A/D converter, connect Vref to Vcc.

```
// atmega328_ad.h
// One must first call ad_ch_init() to initialize the a/d.
// One can repeatedly call ad() after the channel is selected.
// ------------------------------
// Turns off the A/D
void ad_off ()
{ ADCSRA = 0x00;
  return;
} //-----------------------------
//-----------------------------
//Reads current a/d channel, 8 bits
int ad ()
{  return(ADCH);
} //-----------------------------
//-----------------------------
//Selects channel c and reads a/d, 8 bits
int ad_ch (int c)
{    if (c==0) ADMUX = 0x20; // pin 23
     if (c==1) ADMUX = 0x21; // pin 24
     if (c==2) ADMUX = 0x22; // pin 25
     if (c==3) ADMUX = 0x23; // pin 26
     if (c==4) ADMUX = 0x24; // pin 27
     if (c==5) ADMUX = 0x25; // pin 28
```

```
    delay_ms(1);
    return(ADCH);
} // -------------------------------
// -------------------------------
//Initializes a/d, Selects channel c and reads a/d, 8 bits
int ad_ch_init (int c)
{
    DIDR0 = 0x3f;
    ADCSRA = 0xe0;

    if (c==0) ADMUX = 0x20;     // pin 23
    if (c==1) ADMUX = 0x21;     // pin 24
    if (c==2) ADMUX = 0x22;     // pin 25
    if (c==3) ADMUX = 0x23;     // pin 26
    if (c==4) ADMUX = 0x24;     // pin 27
    if (c==5) ADMUX = 0x25;     // pin 28

    unsigned char temp;
    delay_ms(1); temp = ADCH;
    delay_ms(1); temp = ADCH;

    return(temp);
} // -------------------------------
```

## Touch Pad Routines

The touch pad driver subroutines can be a bit tricky to use and the code slightly depends on the specific touch pad, TP, at hand. You might need to closely investigate how the code works to ensure proper operation. The main program should only call tp_read_x() and tp_read_y(). These subroutines return the x and y values respectively. When debugging or trying this for the first time, you may want to only read x values and get that to work first. In fact, in some applications you will only want to read an x value rather than an x and y. In that case, y1 of the TP can be tied to 5v, and y2 of the TP can be tied to 0v. Do not forget that this code uses the A/D converter and as such Vref of the A/D goes to Vcc. Only one side of the TP is pressure sensitive so if in doubt try touching both sides. Since port b has better drive capability than port c, port b is used to drive the 5v on the TP. The 10K pull-down resistors are optional. They pull the voltage to near zero when the TP is not pressed. Without the resistors, the value read from the TP will be near 5v (or 255) when the TP is not pressed. The AVR pins used by this code are: pc1, pc2, pb6, and pb7.

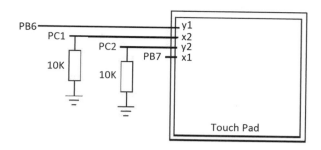

```
// touchpad_driver.h   Oct 2014.
// Minimum value (no press) for the TP is about 37 using 10K pull-down.

// Public subroutines (available for use):
unsigned char tp_read_x ();        // reads the x value of the TP
unsigned char tp_read_y ();        // reads the y value of the TP

// Private subroutines (called only by the public subroutines):
void tp_setup_x2 ();
void tp_setup_y2 ();

/* Physical connection                Reading x            Reading y
pb6, red    , y1                      output, 5v           input, high z
pc1, white, x2 w/10K pull down        input, read x        output, 0v
pc2, green, y2 w/10K pull down        output, 0v           input, read y
pb7, black, x1                        input, high z        output, 5v
*/
#define TP_X1_P        PORTB         // Port assignments
#define TP_Y2_P        PORTC
#define TP_X2_P        PORTC
#define TP_Y1_P        PORTB

#define TP_X1_D        DDRB
#define TP_Y2_D        DDRC
#define TP_X2_D        DDRC
#define TP_Y1_D        DDRB

#define TP_X1       0b10000000       // Pin assignments in port
#define TP_Y2       0b00000100       // AD_Y
#define TP_X2       0b00000010       // AD_X
#define TP_Y1       0b01000000

#define AD_Y    2           // port c a/d channel assignments
#define AD_X    1           // port c a/d channel

#include "ad_328.h"
```

```
//------------------------------------
unsigned char tp_read_x ()   // Gets x value from TP
{   tp_setup_x2();
    unsigned char x = ad_ch_init(AD_X);
    return (x);
}
//------------------------------------
//------------------------------------
unsigned char tp_read_y ()   // Gets y value from TP
{   tp_setup_y2();
    unsigned char y = ad_ch_init(AD_Y);
    return (y);
}
//------------------------------------
//------------------------------------
// Sets up TP to read X2
void tp_setup_x2 ()
{   TP_Y1_D    |=   TP_Y1;     // out
    TP_X2_D    &= ~TP_X2;      // in
    TP_Y2_D    |=   TP_Y2;     // out
    TP_X1_D    &= ~TP_X1;      // in

    TP_Y1_P    |=   TP_Y1;     // 5v
//  TP_X2_P    &= ~TP_X2;      // 0v
    TP_Y2_P    &= ~TP_Y2;      // 0v
//  TP_X1_P    |=   TP_X1;     // 5v
}
// Now one can read x2 from a/d on ch1
//------------------------------------
//------------------------------------
// Sets up TP to read Y2
void tp_setup_y2 ()
{   TP_Y1_D    &= ~TP_Y1;      // in
    TP_X2_D    |=   TP_X2;     // out
    TP_Y2_D    &= ~TP_Y2;      // in
    TP_X1_D    |=   TP_X1;     // out

//  TP_Y1_P    |=   TP_Y1;     // 5v
    TP_X2_P    &= ~TP_X2;      // 0v
```

```
//  TP_Y2_P    &= ~TP_Y2;    // 0v
    TP_X1_P    |=  TP_X1;    // 5v
}
// Now one can read Y2 from a/d on ch2
//————————————————————————————————————
```

# FREQUENCY COUNTER

A frequency counter is a test device that can measure the frequency and period of a periodic wave. The frequency counter demonstrated here is designed to measure periodic waves that are TTL.

If you wanted to measure waves with voltages outside that range, you could attenuate the signal with a voltage divider. The circuit is straight forward. The periodic wave is connected to port b bit 0 of the '328 or '32 and there is a text LCD connected to the AVR. The LCD connections, if not done before, are the hardest part. For connecting a text LCD refer to Chapter 3, "Hardware Interfacing."

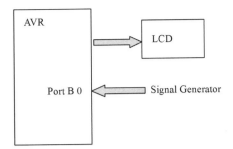

A number of different approaches can be taken with the software. This program uses a polling process. Interrupts could also do the job, but sometimes polling is simpler and easier to debug. The flowchart for the software is:

```
Initialize ports and LCD.
Set the timer to the slowest speed.
Top: Look for a rising edge on the wave.
      Start the timer count at 0.
      Look for the next rising edge on the wave.
      Read the timer count.
      If the timer count is < 1000 then
          Set the timer to a faster speed to get more accuracy.
          Repeat the procedure, Go to Top.
Report the period and frequency (1/period) with proper units.
```

When the timer is started, you need to specify by how much to divide the clock. A clock divided by 1 runs the fastest. Other division values are 8, 64, 256, and 1024. The ATmega328 normally runs at a 1 MHz clock, so divided by 1 produces a timer clock of 1 usec per count. At this rate it would take 65,535 usec to max out the timer count before it overflows. The timer overflow interrupt can be used to catch this event. An example of this along with the timer_328.h code is shown in Chapter 2, "AVR Programming." Frequency can be determined from 1 divided by the period value. This code measures the period and frequency and reports it to the LCD.

```
// ----------------------------------------------------
// atmega328_period.c    Program reports the period and frequency of a wave on pb0.
#include <avr/io.h>
#include "delays_1mhz.h"
#include "lcd_txt_driver328.h"
#include "timer_328.h"
```

```
void wait_for_pb0_to_rise ();
unsigned int read_period (int scale);

int main(void)
{ lcd_init();    // Initialize LCD

  while(1)
   {
    int scale = 1024;    // Try the slowest timer setting.
    float T1;
    T1 = read_period(scale);
    if (T1<1000) // Scale to count faster. Increases resolution.
       { scale = 256;
         T1 = read_period(scale);
       }
    if (T1<1000) // Scale to count faster. Increases resolution.
       { scale = 64;
         T1 = read_period(scale);
       }
    if (T1<1000) // Scale to count faster. Increases resolution.
       { scale = 8;
         T1 = read_period(scale);
       }
    if (T1<1000) // Scale to count faster. Increases resolution.
       { scale = 1;
         T1 = read_period(scale);
       }
    float T;     // T = period
    if (scale==1024) T = T1*1.024;          // msec
    if (scale== 256) T = T1*0.256;          // msec
    if (scale==  64) T = T1*0.064;          // msec
    if (scale==   8) T = T1*8;              // usec
    if (scale==   1) T = T1*1;              // usec

    float f;     // f = frequency
    if (scale==1024) f = 1000/T;            // Hz
    if (scale== 256) f = 1000/T;            // Hz
    if (scale==  64) f = 1000/T;            // Hz
    if (scale==   8) f = 1000000/T;         // Hz
    if (scale==   1) f = 1000000/T;         // Hz

    lcd_cursor_position(LCD_ROW0);
```

```
    if (scale==1024) lcd_ds("T(msec) = $");
    if (scale== 256) lcd_ds("T(msec) = $");
    if (scale==  64) lcd_ds("T(msec) = $");
    if (scale==   8) lcd_ds("T(usec) = $");
    if (scale==   1) lcd_ds("T(usec) = $");
    lcd_dd(T);

    lcd_cursor_position(LCD_ROW1);
    lcd_ds("f( Hz) = $");
    lcd_dd(f);

    delay_ms(500);
  }
} // -------------------------------------------
// ---------------------------------------------
// Returns the period as the number of timer counts at the scale rate.
unsigned int read_period (int scale)
{  timer1_on(scale);                    // Turns on timer
   wait_for_pb0_to_rise();
   timer1_set(0);                       // Set timer to 0
   wait_for_pb0_to_rise();              // Wait for next period to start
   return (timer1_read());              // Reads timer
} // -------------------------------------------
// ---------------------------------------------
// Waits for the value on PB0 to go from low to high.
void wait_for_pb0_to_rise ()
{  while ((PINB&0b01)==0b01); // wait for low
   while ((PINB&0b01)==0b00); // wait for high
} // -------------------------------------------
```

# COIN SORTER

This project works very well for counting coins and adding them up. The most difficult part of the project is the mechanical agitator, and it can be omitted. Coins are placed in the round bin where a geared DC motor rotates a sheet of plastic with holes in it to allow the coins to drop into a slide. The coins roll down a slide and drop into a slot for quarters, nickels, pennies, or dimes—in that order. When the coin passes through a particular slot, it triggers an IR sensor which is connected to an input port on the AVR. An interrupt can be used, but it is easier to poll the 4 sensors to know when a coin passes. The program can add 25 to the sum for quarters, 5 for nickels, 1 for pennies, and 10 for dimes. The output can be displayed on a text LCD.

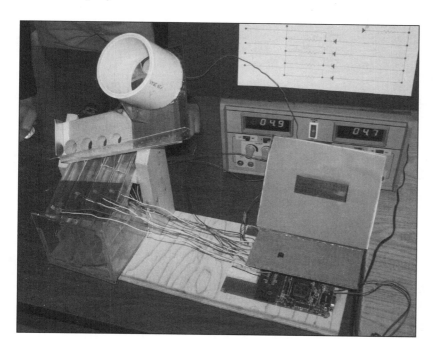

## GUITAR HERO

Guitar Hero is a video game where music is played and the guitar notes scroll down a video screen. The player is to play each note on a guitar as it goes by on the screen. The idea is to build a system that detects the notes and then activates the guitar. The completed project in this picture worked very well, but it would occasionally miss a note. This general idea may be tried on other video games.

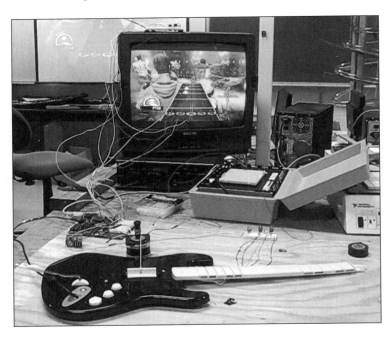

Start with trying to detect the notes as they fall down the screen. To do this, use a photo-resistor and a resistor, R1, in series as shown in the circuit. R1 should be roughly the average value of the photoresistor which is labeled as the sensor. The voltage at V1 is centered at 2.5v. That voltage will swing up and down as the notes go by. The set point resistor and series resistor produces 2.5v at the V2 point. It can be constructed with a single potentiometer. The comparator is to go high each time the note appears and low when there is no note present. The set point potentiometer will be adjusted to allow the output of the comparator to go high and low. The A/D converter could be used to detect a voltage swing from the photoresistor, but the comparator is faster and can easily be adjusted without rebuilding the software.

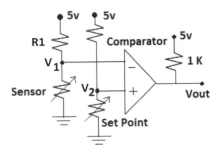

When the sensor and comparator detects a note, the AVR is to play the note located on the guitar arm. To do this, the AVR turns on the associated NPN transistor which is used to switch on the note located on the guitar arm. The guitar arm was opened up and for each note on the arm, the mechanical push switch was replaced with an NPN switch. In addition to the NPN switches, a stepper motor was connected to the AVR. It was used to strum the white paddle shown in the picture.

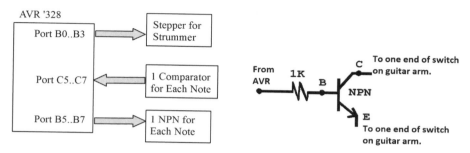

Use this flowchart as a starting point for writing the program in C.

```
Initialize direction of ports B and C.
loop forever
   Read port C
   If bit C5 is active then activate bit B5 for a short time.
   If bit C6 is active then activate bit B6 for a short time.
   If bit C7 is active then activate bit B7 for a short time.
   If any bit C5 to C7 was active then strum the guitar by
      running the stepper motor forward and back.
```

## MORSE CODE TRANSMITTER USING A TOUCH PAD

Morse code is a method of sending letters using dashes and dots. In this project, the AVR transmits the letters as short and long bursts of light from an LED. For example the letter 'A' is represented by a dot, space, dash (short on time, short off time, long on time).

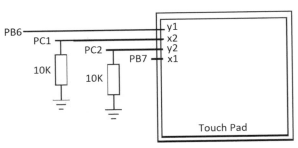

In this picture a touch screen sits on top of a piece of paper listing the letters A to Z. This allows the person to enter letters into the AVR. In the absence of a touch screen, another method would be to use a potentiometer to feed a variable voltage into the A/D converter of the AVR. As the potentiometer rotates, the AVR would see different voltages which would correspond to a letter between A and Z. The letter would be selected when a button is pressed. Using this method, a text LCD would be needed to know what letter is being dialed up.

Sending out the letters to an LED is the easier part of the project. A dash is 3 times longer than a dot and a space is the same duration as a dot. Spaces are used to separate dashes and dots. The program has a single delay subroutine that is the same duration as a dot or space. A dash would need to call it 3 times.

Being a somewhat complicated program, it is best to construct it and test it in stages. First, try sending out letters to the LED without using the touch pad. The touch pad uses the A/D converter, so remember to connect Vref to Vcc. Also, only one side of the touch pad is sensitive.

This code reads the TP for a letter and transmits the Morse code using an LED. It includes three other files. The program uses an optional text LCD to provide feedback as to what values are being read from the touch pad. It is possible to receive feedback data using the USART on the AVR connected to a terminal window on the personal computer.

```c
// atmega328_morse_code.c

#include <avr/io.h>
#include "delays_1mhz.h"
#include "lcd_txt_driver328.h"
#include "touchpad_driver.h"

#define debug 2    //1, 2 or 3 to enable debugging output. 0 to disable debugging.

void dot();
void dash();
void delay();
char tp_verify_char();
char tp_read_char ();

int main(void)
{
DDRB |= 0b01;          // For LED to see morse code.
lcd_init();

if (debug==1)
   do { lcd_home();
        lcd_dd(tp_read_x()); lcd_da(' ');
        lcd_dd(tp_read_y());
        delay_ms(2000);
      } while (1);

char letter = 'B';
do {
      letter = tp_verify_char();
    //letter = tp_read_char();    // Used here for testing purposes
      if (debug==2) { lcd_home(); lcd_da(letter); }
      if (letter=='A') { dot(); dash(); }
      if (letter=='B') {dash();  dot();  dot();  dot();  }
      if (letter=='C') {dash();  dot(); dash();  dot();  }

      delay_ms(1000);
    } while (1);
}
// ---------------------------------
void dot()
{   PORTB = 0b01; delay();
    PORTB = 0b00; delay();
}
// ---------------------------------
```

```
// ------------------------------------------
void dash()
{   PORTB = 0b01; delay(); delay(); delay();
    PORTB = 0b00; delay();
}
// ------------------------------------------
// ------------------------------------------
void delay()    // 1/4 sec delay
{   for (int i=0;i<13156;i++);
}
// ------------------------------------------
// ------------------------------------------
char tp_verify_char()
{   // A '$' is read from the touchpad, TP, if it is not pressed.
    // This returns a letter other than a '$' from the TP.
    // It rejects erratic values by requiring 3 values in a row to be the same.
     char selected = 0, c1='$',c2='$',c3='$';
    do {
            c1 = c2;                    // shift letters down
            c2 = c3;
            c3 = tp_read_char();        // get new character

            if (c1 == c2)               // must get 3 in a row of the same letter.
               if (c2 == c3)
                  if (c3!='$') (selected = 1);
          } while (!selected);
        return (c3);
}
// ------------------------------------------
// ------------------------------------------
/* The touchpad, tp, has a paper behind it with 4 lines as shown.
      7 columns of letters per line.
      24 values out of 255 (from A/D) per column in the x direction.
      40 values out of 255 (from A/D) per row     in the y direction.
      x boundaries (c1 to c6) = 187, 165, 143, 121, 99, 75
      y boundaries (r1 to r4) = 160, 120, 80, 40

   a b c d e f g
   h i j k l m n
   o p q r s t u
   v w x y z $ $
*/
```

```
char tp_read_char ()
{ int const c1 = 187;
  int const c2 = 165;
  int const c3 = 143;
  int const c4 = 121;
  int const c5 =  99;
  int const c6 =  75;

  int const r1 = 160;
  int const r2 = 120;
  int const r3 =  80;
  int const r4 =  40;

  char x = tp_read_x();
  char y = tp_read_y();

  if(debug==3){lcd_home();lcd_dd(x);lcd_da(',');lcd_dd(y);delay_ms(1000);}

  if (y>r1)
  {   if (x>c1) return('A');
      if (x>c2) return('B');
      if (x>c3) return('C');
      if (x>c4) return('D');
      if (x>c5) return('E');
      if (x>c6) return('F');
                return('G');
  }
  if (y>r2)
  {   if (x>c1) return('H');
      if (x>c2) return('I');
      if (x>c3) return('J');
      if (x>c4) return('K');
      if (x>c5) return('L');
      if (x>c6) return('M');
      return('N');
  }
  if (y>r3)
  {   if (x>c1) return('O');
      if (x>c2) return('P');
      if (x>c3) return('Q');
      if (x>c4) return('R');
      if (x>c5) return('S');
      if (x>c6) return('T');
      return('U');
  }
```

```
   // Otherwise bottom row...
   { if (x>c1) return('V');
     if (x>c2) return('W');
     if (x>c3) return('X');
     if (x>c4) return('Y');
     if (x>c5) return('Z');
     if (x>c6) return(' ');
     return('$');
   }
}
// ----------------------------------
```

## SPEED MEASUREMENT USING THE TIMER

The speed of an object, such as a toy car on a track or any other object, can be measured using an IR emitter and detector along with the AVR's timer. Chapter 3, "Hardware Interfacing," shows how to build an IR circuit that normally sends out a high signal, but when an object blocks the infrared light it sends out a low signal. The output of this circuit will go to one of the I/O pins on the AVR, such as pb0.

Construct two IR circuits on the car track a specific distance apart, say 1 ft. The speed of the object going down the track is then computed from distance between sensors / time between sensors. Use the timer to measure the time between sensors. The timer has a maximum count of 65,535, and at the maximum counting rate of 1 count per usec (when the frequency is 1 MHz), it takes only 65 msec to overflow. You will most likely want to slow down the count rate by a factor of 8, 64, 256, or 1024. Refer to the timer counter and example code in Chapter 2, "AVR Programming." Use this flowchart for the program that will find the time for an event to occur.

```
Read (polling) the first IR until it goes low
Turn on timer and set timer to zero
Read (polling) the second IR until it goes low
Read the timer.
Compute speed = distance / time.
Report speed to LCD with appropriate units.
```

For example, if the distance between IR sensors is 2ft and the timer returns 166 msec then the speed is equal to 2 ft/ 166 msec = 0.01205 ft/msec = 12.05 ft/sec. This math can easily be done in C rather than Assembly. If the variable to represent speed is an integer, then the number will be truncated yielding a 12 in the example. If you went with mm for units, then the speed would be 306 mm/s which has 3 significant figures rather than 2.

The next step in the project would be to connect a text screen to the system. For this information refer to text screens in Chapter 3, "Hardware Interfacing."

## GPS REPORTING LATITUDE, LONGITUDE, AND TIME

This project demonstrates how to collect data from a GPS sensor, parse the data, and report the information on a text LCD. The GPS sensor used here, PMB-648, uses 3.3v to 5v for power. It sends out a character string containing data separated by commas at 4800 baud. For more information, refer to the GPS sensor in Chapter 3, "Hardware Interfacing." The '328 has its USART0 receive on pd0; therefore, pd0 must be connected to the output of the GPS.

The text LCD used here has an SPI interface. On the AVR side, the SPI is implemented in software on port d. The LCD may not use pd0, because that is being used by the GPS. The LCD used here requires 3.3v for power. For this reason, the '328 will also use 3.3v. If your GPS only operates at 5v then connect the output of the GPS to a voltage divider that reduces its 5v highs to 3.3v highs. The voltage divider can be made with a 2K resistor in series with a 3K resistor or a potentiometer.

The program uses classes, which is supported in AVR Studio version 6 and later. Here is the flowchart for the program.

```
Initialize ports, LCD, and USART0.
Wait until a "$GPRMC" is received.
    Then count commas until token of interest appears.
    Record or display token of interest.
```

The program just displays the values on the LCD, but you might want to log the data. It is easy to store the values of interest in EEPROM, as shown in Chapter 2, "AVR Programming." To upload the EEPROM to the PC, use AVR Studio and click on Tools > AVR Programming. Under Memories select Read EEPROM. Once the data is on the PC, you will have to extract the data from the file.

Another method for uploading data to the PC is to use a terminal window on the PC. Use the USART0 transmit pin on the AVR to transmit the data to a serial input pin on the programming cable. There is an optional terminal window for AVR Studio that you can download and use. There are programs and applications that can take a collection of latitude and longitude points and plot the course traveled.

```
/* atmega328_gps_lcd16x2_cpp.cpp,
 This program collects data from the GPS and displays it on the SPI text LCD.
 This uses the SPI text LCD screen at 3.3v.
```

```
Use pd7 (not pd0) as the MOSI pin for the text SPI LCD.
Pd0 of the '328 is to receive the GPS message at 4800 Baud.
*/

#include <avr/io.h>
#include "lcd_16x2spi_cpp.h" //listed in chapter 2 under SPI Implementation.
#include "serial_328cpp.h"   //listed in chapter 2 under USART0 for the '328

int main(void)
{ // The number is the item or token number in the GPRMC string.
  const int GPS_UTC        = 1;    //Universal time coordinated.
  const int GPS_LATITUDE   = 3;
  const int GPS_LONGITUDE  = 5;
  const int GPS_SPEED      = 7;
  const int GPS_COURSE     = 8;
  const int GPS_DATE       = 9;

  #define BUTTON 0b01           // Used to display location or time.

  class lcd16x2 lcd;
  class serial328 gps;

  gps.init_rx(BAUD4800);    // Initialize USART0 for GPS
  char s[100];              // Records message from GPS
  DDRB &= ~BUTTON;          // Switch is on pb0

  while (1)
  { //  lcd.da('a'); while (1);              // Used to see if LCD works
    do {} while ( gps.rx() != '$');         // Wait for start of string.

    s[0] = gps.rx();    // Get the first 5 characters after $
    s[1] = gps.rx();
    s[2] = gps.rx();
    s[3] = gps.rx();
    s[4] = gps.rx();

    if (s[0]=='G')      // Is it a GPRMC message?
    if (s[1]=='P')
    if (s[2]=='R')
    if (s[3]=='M')
    if (s[4]=='C')
      {
        int item;                        // Token number to display
        int count;                       // counts items (tokens) in string
        int i;                           // counts chars in string

        for (int i=5;i<100;i++)   s[i] = gps.rx(); //Grab string of 60 characters.
        lcd.clear();                     // clear LCD
```

```
       int button = PINB & BUTTON;
       if(button)                          // Report location
       { item=GPS_LATITUDE;
         count=i=0;
         lcd.line1(0);
         lcd.ds("Lat=$");
         do { if (s[i++]==',') count++;
             } while (count<item);          // loop until count=item
         do { lcd.da(s[i]);
             } while (s[++i]!=',');         // Display this item

         item=GPS_LONGITUDE;
         count=i=0;
         lcd.line2(0);
         lcd.ds("Lon=$");
         do { if (s[i++]==',') count++;
             } while (count<item);          // loop until count=item
         do { lcd.da(s[i]);
             } while (s[++i]!=',');         // Display this item
       } // end if button

       if(!button)                          // Display time.
       { item=GPS_UTC;
         count=i=0;
         lcd.line1(0);
         lcd.ds("UTC=$");
         do { if (s[i++]==',') count++;
             } while (count<item);           // loop until count=item
         do { lcd.da(s[i]);
             } while (s[++i]!=',');         // Display this item

         item=GPS_SPEED;
         count=i=0;
         lcd.line2(0);
         lcd.ds("Speed=$");
         do { if (s[i++]==',') count++;
             } while (count<item);          // loop until count=item
         do { lcd.da(s[i]);
             } while (s[++i]!=',');         // Display this item
       } // end if !button
     } // end if
   } // end while
} // end main
```

# MEASURING AN INCLINE

In some cases you want to measure the pitch of a roof, incline of the road, the grade of a bulldozer. This can be done with an AVR, text screen, and a pendulum connected to a potentiometer as shown in the picture.

When this device is on level ground, the potentiometer is at an angle of 0 degrees. The pendulum will always point to the nadir. As the inclination changes so does the potentiometer angle due to the pendulum. The potentiometer has 3 wires. The outer 2 go to 5v and ground while the middle wire goes to one of the A/D channels. Also connect a text LCD screen to the AVR. The subroutines for the LCD are found in Chapter 3, "Hardware Interfacing." This LCD uses all of port d, pc4, and pc5.

After connecting the hardware, display the A/D values for angles 0 and 90 degrees and call them ad0 and ad90 respectively. Use these values in your final program to convert an A/D value to an angle as shown by the equation. In your final program, have the AVR measure the A/D value, compute the measured angle, and display this measured angle as shown in the program.

Measured Angle = 90 × (Measured A/D value - ad0)/(ad90-ad0)

```
// – – – – – – – – – – – – – – – – – – – – – – – – – – – – – – – – –
// atmega328_ad_angle.c
// Equation converts voltage from my pot to an angle. Each pot will be different!
// Connect pot middle wiper of pot to A/D ch0. Outer 2 legs to 5v and 0v.

#include <avr/io.h>
#include "delays_1mhz.h"
#include "ad_328.h"
#include "lcd_txt_driver328.h"

int main (void)
{   lcd_init();
    do {   // Read a/d and convert to angle. Your equation will differ.
        int angle = 1*ad_ch_init(0)-110;
        if (angle<0) { lcd_da('-'); angle = -angle; }
            else lcd_da('+');
        lcd_dd(angle);                          // display angle
        delay_ms(200);
        lcd_home();
        } while (1);
}
// – – – – – – – – – – – – – – – – – – – – – – – – – – – – – – – – –
```

# ROTATING LED MESSAGE

The rotating message project typically has 8 LEDs mounted on an arm which is fitted onto the shaft of a DC motor. Wood or even stiff cardboard works well for the arm as it is easily worked. As the motor spins, the LEDs are timed to light up at the correct position such that the observer will see a message that appears stationary. When the arm spins, be careful not to go too fast, cause it to vibrate, or touch the spinning arm.

In the next two pictures, a 3D printer was used to make the base along with 3 arms that snap into the base. The base is then mounted onto the motor shaft. The motor is held fast by a large wood clamp. Each arm contains 8 LEDs of a single color. One arm is for red, one for yellow, and one for blue. The 24 LEDs require 3 available ports and for that reason an ATmega32 was chosen.

The block diagram shows how the hardware connects to the AVR. The picture is of a rotating LED project spelling out some words.

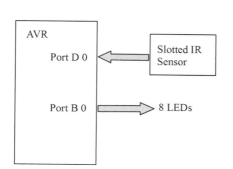
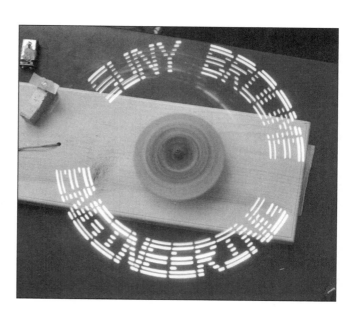

Find the parts needed and build the mechanics first. Getting the appropriate parts can actually be the hardest part. Do not make the arm longer than about 10 inches. See that the motor can rotate at a slow speed without vibration. Be sure to have all parts secured. The AVR board and battery will need to be centered and secured onto the rotating base to minimize vibration. Check that the AVR can light up the 8 LEDs. The AVR will use a slotted IR sensor or similar to know when the arm is at the 12 o'clock position. This will be used for timing purposes. To check the sensor, use your hand to rotate the arm into the sensor. Write a small program to have the AVR detect the sensor and then turn on the LEDs for about 1 second. When the arm is not in the sensor, the LEDs will turn off.

The next part will take a lot of trial and error. We are now ready to display something. Start with a simple number like 1 to display. Write a program that will detect the sensor and when detected have the AVR light all 8 LEDs for a fraction of time. If the arm revolves 5 times per second then one circle is 1/5. A 1 should be about 1/100 of the circle, so the fraction of time would be on the order of 1/500 second or 2 ms. Write a delay for 1 or 2 ms. See how this performs. The flowchart would look like this:

```
Initialize ports
Infinite loop to display a 1:
    Wait for IR sensor to become active
    Light all 8 LEDs
    Delay 1 ms
    Turn off all 8 LEDs
```

Once this works then you can investigate more creative words and images to display. This flowchart displays the number 9, albeit not a fancy one. The delay may need to be adjusted.

```
Initialize ports
Infinite loop to display a 9:
    Wait for IR sensor active to indicate the starting position.
    Light LEDs 0b00001111, Delay 1 ms
    Light LEDs 0b00001001, Delay 1 ms
    Light LEDs 0b00001001, Delay 1 ms
    Light LEDs 0b11111111, Delay 1 ms
    Turn off all 8 LEDs
```

## Sketch Pad Using a Graphics LCD and Potentiometer

This project draws lines and shapes on a graphics screen using 2 potentiometers for input. The starting point is selecting a graphics LCD screen, then connecting it to the AVR and writing code for it. This can involve a bit of work. Chapter 3 shows how to interface a graphics screen. It uses all of port d for data and pb0 to pb5 for control signals. Most other graphics screens would be interfaced in a similar fashion. Use an ATmega328 or

ATmega32, because the ATtiny13 does not have enough I/O. After the LCD is connected, write and debug the code to get a few pixels to light up.

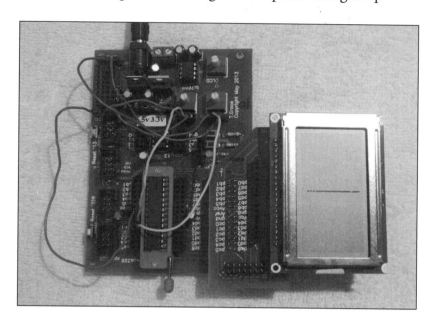

Once this works, add 2 potentiometers. One for selecting the x position and one for selecting the y position. The outside legs of the potentiometers go to 5v and ground. Each middle leg (wiper) will go to a separate A/D channel. The values of the A/D will determine what pixel (x,y) to light up. Try to keep the wires going from the potentiometer to the A/D as short as possible and shielded from electronics that produce electromagnetic radiation. The A/D can easily pick up stray signals. In fact, you may want to see if you can pick them up and display them on the screen. What follows is a general flowchart and program.

```
// – – – – – – – – – – – – – – – – – – – – – – – – – – – – – – – – – – – – –
//Include LCD and A/D files
//Initialize LCD
//Infinite loop:
//      Read x from A/D channel 1
//      Read y from A/D channel 2
//      x = x/2    // Scale to fit on screen
//      y = y/4    // Scale to fit on screen
//      display pixel (x,y)

//Atmega328_sketchpad_potentiometer.c
```

```
#include <avr/io.h>
#include "delays_1mhz.h"
#include "atmega328_ad.h"
#include "lcd_64gs12L4_driver328.h"

int main(void)
{  lcd64_init();               // Required
   lcd64_clr();                // Clear screen
   ad_ch_init(1);
   while (1)
     { int x = ad_ch(1);
       int y = ad_ch(2);
       lcd64_xy(x/2,y/4,1);
     }
   return 0;
} // - - - - - - - - - - - - - - - - - - - - - - - - - - - - - - - - - -
```

You can add a button that will clear the screen when pushed. A cool idea would be to add an accelerometer as described in Chapter 3, "Hardware Interfacing." When the project is gently shaken, the accelerometer will detect this and tell the AVR to clear the screen. The accelerometer would be connected to one of the A/D channels.

## Sketch Pad Using a Graphics LCD and Touch Pad

In this project, a touch pad is placed on top of a graphics LCD. As you press on the touch pad, the underlying pixels light up. This is very similar to a signature pad for a credit card. Refer to the project on Graphics LCD and Potentiometer as a starting point. You may want to try that project first. For this project, instead of adding the potentiometers, connect the touch pad to the A/D. Refer to the sections and examples on the touch pad. The flowchart for the main program would be something like this. Notice how similar the flowchart and program is to the prior example. Also notice that the ports used by the graphics screen do not overlap with those used by the touch pad.

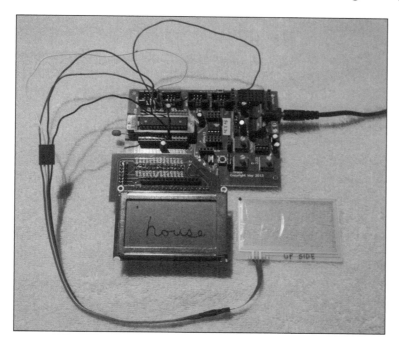

```
// -----------------------------------------
//Include LCD, Touch Pad and A/D files
//Initialize LCD
//Infinite loop:
//    Read xv from Touch Pad
//    Read yv from Touch Pad
//    x = xv*128/256    // Scale to fit on screen
//    y = yv* 64/256    // Scale to fit on screen
//    display pixel (x,y)
//
//    atmega328_sketchpad_touchpad.c
#include <avr/io.h>
#include "delays_1mhz.h"              // Turn off optimization
// #include "atmega328_ad.h"          // Do not forget to connect Vref.
#include "lcd_64gs12L4_driver328.h"
#include "touchpad_driver.h"          // Includes the file atmega328_ad.h

int main(void)
{   lcd64_init();                // Required
    lcd64_clr();                 // Clear screen

    while (1)
     { int x = tp_read_x();
       int y = tp_read_y();
```

```
    lcd64_xy (x/2,y/4,1);
  }
 return 0;
} // -----------------------------------------
```

## Tic-Tac-Toe Using a Graphics LCD

This fun idea builds on the prior project which is to get a touch pad and graphics screen to work together. In addition to displaying individual pixels, the program will also need to draw lines for the board which can be done by calling this subroutine found in the graphics include file.

```
lcd64_line(41, 0, 41,63,1);    //    Vertical line for board
lcd64_line(82, 0, 82,63,1);    //    Vertical
lcd64_line( 0,21,127,21,1);    // Horizontal
lcd64_line( 0,42,127,42,1);    // Horizontal
```

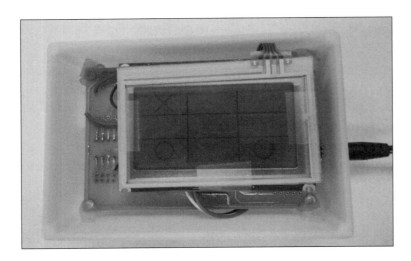

A person will play against the computer. The program to do this can be long and involved. A good approach would be to write and debug the tic-tac-toe logic on a personal computer using a C compiler before trying to move the code to the AVR. It is easier to debug the program logic on the PC. After the program is debugged, move the program to AVR Studio. The flowchart will look something like this. A 3D printer can make a nice case for the finished project.

```
done = 0, turn = 'x' for computer, draw board.
loop until done = 1
    if turn='x' then
            Check to see if computer can win and go there.
        else  Check to see if person can win and block it.
        else  Get a random number (1 to 9) from timer and
            draw X on board if space is free, otherwise repeat.
    if turn='o' then read touch pad to get location to draw O.
    if any row is all X or O, then done = 1.
    if any column is all X or O, then done = 1.
    if any diagonal is all X or O, then done = 1.
    if all spaces occupied then done = 1.
    if turn='x' then turn='o' else turn='x'.
```

## CONTROLLING A SERVO MOTOR

This project shows how to control a servo motor using an ATtiny13 in Assembly. Using an ATmega328 or ATmega32 would be identical with the exception of calculating the delay times. This is the easiest project involving a servo, so you may want to try this first. Most servo motors have a range of motion −90 degrees to +90 degrees. The angle of the servo is determined by the high time of a 20 msec periodic wave as shown in the table. On the servo, connect the red wire and black wire to 5v and 0v respectively. The other wire, typically white or yellow, connects to the periodic wave produced by the AVR.

| High Time | Angle of Servo |
|---|---|
| 1.25 msec | −90 |
| 1.50 | 0 |
| 1.75 | +90 |

The high times shown in the preceding table may vary from servo to servo, so you will need to experiment with the delay time in order to get the desired angle. The default clock speed of the '13 is 1,200,000 cycles/sec while the '328 and '32 is 1,000,000 cycles/ sec. To position the servo at 0 degrees, a high time of 1.5 msec is desired. Since the period is 20 msec, that means the low time is 18.5 msec. The period, 20 msec, does not have to be perfect and has some wiggle room. That means we can keep the low time at 18.5 msec for all cases. How many cycles and loops constitute 18.5 msec for the ATtiny13? The delay loop in the program uses a sbiw and brne instruction. Both instructions take 2 cycles. The loop takes 4 cycles per loop.

18500 usec × (1,200,000 cycles/ 1,000,000 usec) = 22,200 cycles

22,200 cycles × 1 loop/ (4 cycles) = 5,550 loops

Similar calculations were performed to arrive at the number of loops for +/−90 degrees. After the calculation, the numbers were tested and modified to better match the actual servo at hand. The servo used here is from Jameco Electronics (part #2150416).

```
#define SERVO_CW     200     ;delay to reach full CW    -90,   1.25 msec
#define SERVO_CCW    545     ;delay to reach full CCW   +90,   1.75 msec
```

In the main program, do not forget to set the data direction register. The main program drives the servo fully CW for 2 seconds then fully CCW for 2 seconds. 100 loops of 1 period (20 msec) equals 2 seconds. The main program passes to the subroutine delay_hightime, the time that the servo pulse needs to be high for. The register z contains this delay value.

```
; — — — — — — — — — — — — — — — — — — — —
/* attiny13_servo2.asm
Program runs the servo full CW then CCW for 2 sec. each.
The delay values work for the small blue servo.
*/
#define port           portb
#define port_ddr       ddrb
#define servo_l        r24      ;a low to servo
#define servo_h        r25      ;a high to servo
#define SERVO_CW       200      ;delay to reach full CW    -90
#define SERVO_CCW      545      ;delay to reach full CCW   +90

.org    0x0000
   ldi    servo_l,0b00
   ldi    servo_h,0b01
   out    port_ddr,servo_h    ;set pb0 to output for servo.
```

```
main1:
    ldi    r16,100             ;Full clockwise for 2 sec.
short:
    ldi    zl,SERVO_CW %256
    ldi    zh,SERVO_CW /256
    out    port,servo_h
    rcall delay_hightime       ;Narrow pulse to go CW
    out    port,servo_l
    rcall delay_18p5ms
    dec    r16
    brne   short

    ldi    r16,100             ;Full counter-clockwise for 2 sec.
long:
    ldi    zl,SERVO_CCW %256
    ldi    zh,SERVO_CCW /256
    out    port,servo_h
    rcall delay_hightime       ;Wide pulse to go CCW
    out    port,servo_l
    rcall delay_18p5ms
    dec    r16
    brne   long

    rjmp   main1
; ------------------------------
; ------------------------------
; hightime usec * 1.2cycles/usec * 1 loop/ (4 cycles) = z = number of loops
delay_hightime:
;    ldi    zl,<value> %256
;    ldi    zh,<value> /256
delay_hightime1:
    sbiw z,1
    brne delay_hightime1
    ret
; ------------------------------
; ------------------------------
; 18500 usec * 1.2cycles/usec * 1 loop/ (4 cycles) = 5550
delay_18p5ms:
    ldi    zl,5550 %256
    ldi    zh,5550 /256
delay_18p5ms1:
    sbiw z,1
    brne delay_18p5ms1
    ret
; ------------------------------
```

# THE USELESS MACHINE USING A SERVO

The useless machine as shown by the picture has a servo, ATtiny13, and a large switch. When a person turns the switch on, the '13 will tell the servo to turn the switch off to annoy the human. Interesting, but feckless.

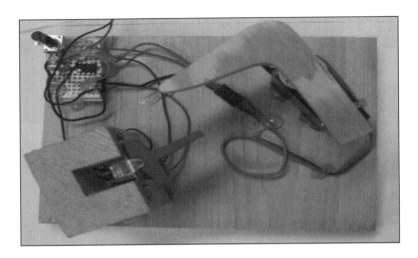

The servo has its own power supply with a common ground. A common ground is connected to all other grounds. The control line of the servo is a square wave with a period of 20 ms. The high time of the square wave varies from 1.25 to 1.75 ms. The high time controls the position of the servo. The program is to maintain a period of 20 ms for the square wave. This number has some allowable play thus the low time can be fixed to 18.5 ms regardless of the high time variation.

In the program, the first thing to do is to list the subroutine prototypes, set up the DDRB and then enter an infinite loop. In the loop, read the switch and isolate the one bit that pertains to that pin. If the switch is in the off position, the servo will retract for 2 seconds. To do this, the AVR will output a high time of 1.75 ms. This will keep the servo retracted and the switch in the off position. If the switch is in the on position, the servo will extend for 2 seconds to turn the switch off. To extend, the AVR will output a high time of 1.25 ms. One period is 20 ms, so 100 periods is 2 seconds. These delay routines are designed for an ATtiny13 running at 1.2 MHz.

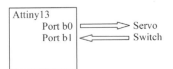

```
// useless_machine.c
#include <avr/io.h>
void delay18_5 ();        //delay of 18.5ms
void delay1_25 ();        //delay of 1.25ms
void delay1_75 ();        //delay of 1.75ms


int main(void)           // A switch is connected to bit 1 of portb.
{   DDRB = 0b01;          // Servo is connected to bit 0 of portb.
    while(1)
      { int x = PINB;             // Read switch.
        x &= 0b00000010;          // Isolate switch.

      for (int i=0;i<100;i++) // Repeat action for 2 sec.
        {
            PORTB |= 0b00000001;       // Set servo line high.
            if (x)   delay1_75();      // Move switch.
                else delay1_25();      // Retract servo arm.
            PORTB &= 0b11111110;       // Set servo line low.
            delay18_5();
        }
    }
}
void delay18_5 () {   for (int i=0;i<1200;i++); }
void delay1_25 () {   for (int i=0;i<  81;i++); }
void delay1_75 () {   for (int i=0;i< 113;i++); }
```

## SUN LOCATOR USING A SERVO

The sun locator project uses an ATtiny13, its A/D, and photoresistor mounted on a servo to search for the brightest spot in a 180-degree arc. The servo moves the photoresistor in an arc searching for the brightest spot. The programming is done in Assembly. First, attain a servo and determine what delay values are needed to make it move fully clockwise and fully counter-clockwise. The delay values here work for the servo from Jameco Electronics (part #2150416). If you have not worked with a servo before you may want to try the first project on servos called Controlling a Servo Motor.

Connect the photoresistor circuit as shown to the channel 1 A/D on the '13. The photoresistor, R2, has less resistance in greater light. When the light increases, the voltage going to the A/D also increases. Select R1 as the average value of the photoresistor.

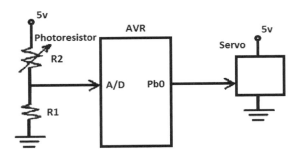

The program starts out by moving the servo fully CCW for 2 seconds. The servo then slowly moves CW sampling the light intensity via the photoresistor and A/D converter. When a new high value is found, the AVR remembers the intensity and position. At the end of the servo range, the servo moves back to where the high value was detected. This concept can be used for other parameters like measuring distances to objects in an arc. To measure distance, you can use the IR distance sensor or ultrasonic distance sensor covered in Chapter 3, "Hardware Interfacing." Another variation on this project would be to connect a graphics screen to the AVR and plot the light or distance versus position on the screen. In this case, you would need to use an ATmega328 or ATmega32 due to the I/O required by the LCD. See what resources are needed by the LCD first, then add the A/D, and lastly add the servo, because the servo can go pretty much on any I/O pin. This code looks for the position where there is greatest light.

```
/* attiny13_servo2.asm     18 Sept 2014
Program runs the servo full CCW for 2 sec. Then moves CW looking for the highest value
on ad1 (pb2). After scanning the angular range, servo moves to angular position with
highest ad1 value.
*/
#define port            portb
#define port_ddr        ddrb

#define ad1             r21      ;Value read by a/d on ch1
#define repeat          r22      ;Number of times to repeat servo wave.
#define ad1_max         r23      ;maximum value read by a/d on ch1.
#define servo_l         r24      ;a low to servo
#define servo_h         r25      ;a high to servo

#define SERVO_CW        200      ;delay to reach full CW
#define SERVO_CCW       545      ;delay to reach full CCW
.org    0x0000
    ldi    servo_l,0b00
    ldi    servo_h,0b01
    out    port_ddr,servo_h      ;set pb0 to output for servo.
    rcall attiny13_ad1_init      ;set up ch1 (pb2) for a/d
main1:
    ldi    yl,SERVO_CCW %256
    ldi    yh,SERVO_CCW /256
    ldi    ad1_max,0             ;initialize maximum to 0
    ldi    repeat,100            ;Allow 2 sec to reach starting position.
main3:
main2:
    mov    zl,yl
    mov    zh,yh
    out    port,servo_h
    rcall delay_hightime
    out    port,servo_l
    rcall delay_18p5ms
    dec    repeat
    brne   main2

    ldi     repeat,1
```

```
        in      ad1,adch            ; Read the value of the photoresistor
        cp      ad1,ad1_max         ; Is this the brightest value so far?
        brlo    main4
        mov     ad1_max,ad1         ; a new maximum found
        mov     xl,yl               ; remember position for maximum
        mov     xh,yh
main4:
        sbiw    y,1                 ; Did servo reach the end of travel?
        cpi     yl,SERVO_CW %256
        brne    main3
        cpi     yh,SERVO_CW /256
        brne    main3               ; Continue rotating
main9:                              ; loop forever at the position where maximum occurs
        mov     zl,xl               ; x is the position for maximum
        mov     zh,xh
        out     port,servo_h
        rcall   delay_hightime
        out     port,servo_l
        rcall   delay_18p5ms
        rjmp    main9
; ------------------------------
; ------------------------------
; hightime usec * 1.2cycles/usec * 1 loop/ (4 cycles) = z = number of loops
delay_hightime:
;       ldi     zl,<value> %256
;       ldi     zh,<value> /256
delay_hightime1:
        sbiw z,1
        brne delay_hightime1
        ret
; ------------------------------
; ------------------------------
; 18500 usec * 1.2cycles/usec * 1 loop/ (4 cycles) = 5550
delay_18p5ms:
        ldi     zl,5550 %256
        ldi     zh,5550 /256
delay_18p5ms1:
        sbiw z,1
        brne delay_18p5ms1
        ret
; ------------------------------
```

```
; ─────────────────────────
attiny13_ad1_init:              ;for a/d ch1 setup on pb2 of attiny13
    push   r16
    ldi    r16,0x3c
    out    didr0,r16            ;digital input disable
    ldi    r16,0x21
    out    admux,r16            ;8 bits channel 1
    ldi    r16,0xe0
    out    adcsra,r16           ;continuously update
    pop    r16
    ret
; ─────────────────────────
```

## SERVO WITH TIMER INTERRUPT EXAMPLE

This example uses the built-in timer on the ATmega32 to control a servo on PORTB. The DDRB needs to be set to output. The general idea of the program is to move the servo back and forth. The timer interrupt is used to keep track of the waveform time rather than using a dedicated delay subroutine. The general flowchart is as follows.

```
For 200 msec move toward the -90 position by doing this:
Loop 10 times, each loop is 20 msec long. In each loop do this:
        set the servo control line high, wait for 1250 usec.
        set the servo control line low, wait (20,000-1250) usec.

For 200 msec move toward the +90 position by doing this:
Loop 10 times, each loop is 20 msec long. In each loop do this:
        set the servo control line high, wait for 1750 usec.
        set the servo control line low, wait (20,000-1750) usec.
```

The program sets the servo line to the appropriate state (high or low) and then sets the timer for the required time. To make the timer wait N microseconds, set the timer to (0xffff-N) since the timer clock counts up once every microsecond. This is because the default clock on the '328 and '32 is 1 MHz.

A loop that does nothing is entered waiting for the timer to expire. When the time is up, a timer interrupt is generated which runs the ISR. The ISR turns off the timer, sets the global variable done to 1, and returns. Upon returning to the main program, the variable done kicks the program out of the while loop. The variable done needs to be volatile so that the compiler does not convert while(!done) to while(1), which is an infinite loop. If the qualifier is not volatile, the compiler assumes that the variable done never changes from a zero. If you are using the '328, change TIMSK to TIMSK1. The main program is listed.

```c
// atmega32_servo.c
#include <avr/io.h>
#include <avr/interrupt.h>
#include "timer_32.h"        // Code is listed in chapter 2 under timer counter.
volatile int done = 0;

int main(void)
{ DDRB = 0xff;              // Output to servo. Servo can be on any pin of port b.
  timer1_off();            // timer off
  timer_isr_enable();  // enable interrupts

  while(1)
    {
      for (int i=0;i<10;i++)         // Move toward position 1 for 200 msec.
        { PORTB = 0xff; timer1_set(0xffff-1250);
                        done=0; timer1_on(1); while(!done);
          PORTB = 0b00; timer1_set(0xffff-(20000-1250));
                        done=0; timer1_on(1); while(!done);
        }
      for (int i=0;i<10;i++)         // Move toward position 2 for 200 msec.
        { PORTB = 0xff; timer1_set(0xffff-1750);
                        done=0; timer1_on(1); while(!done);
          PORTB = 0b00; timer1_set(0xffff-(20000-1750));
                        done=0; timer1_on(1); while(!done);
        }
    }
}
```

It is interesting to note that the 200 msec may or may not be enough time for the servo motor to reach the desired position. The time it takes to reach the desired position depends on the servo speed. Remember that the waveform sent to the servo determines position (not speed). One waveform is only 1/50 second which is usually not enough time to reach the position. Here is the ISR for the prior program. The timer_32.h code is listed in Chapter 2 under "Timer Counter."

```c
// ------------------------------------------------
// Interrupt service routine. Makes variable done = true.
ISR(TIMER1_OVF_vect)
{   done = 1;
    timer1_off();
    return;
}
// ------------------------------------------------
```

# MASS BALANCE USING A STEPPER

This balance worked very well and was able to measure masses with a gram. To determine the mass, the AVR uses a stepper motor to move a carriage back and forth on a pivoted arm. When the arm becomes level, the AVR computer can determine the mass from the number of steps the stepper took starting from the zero position. To get the conversion between steps taken and mass in grams, you need to put a known mass on the scale and count the number of steps to level it. This ratio of steps per gram is the conversion factor used in the program.

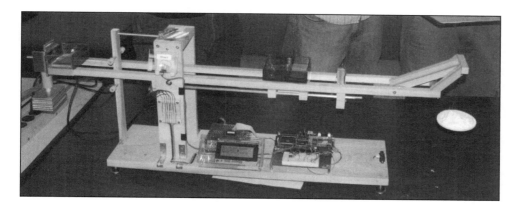

In addition to a '328 or '32, the project utilizes a text LCD, stepper motor, stepper motor driver circuit, and an IR distance sensor. As shown in the picture, the IR distance sensor points up at the end of the arm and reports back to the AVR the distance to the arm. When the arm is at the correct distance, the arm is level and the mass is reported. Initially the carriage is at the far left, the unknown mass is on the scale making the arm fall down. The AVR repeatedly steps the motor while checking the distance. The hardware construction is more difficult than the software.

This particular scale was made from an old inkjet printer. The printer's stepper and carriage were kept intact, which saved a lot of time in construction. If you do not have an IR distance sensor, you can do what the Soviets did a long time ago on their rockets. They made a cage with a wooden top and bottom. Running vertically on the perimeter of the cage were numerous wires with a metal ball inside. As the rocket tipped, the metal ball would go to the lowest side and conduct on 2 wires. This same approach can be used here to know which way the arm is tilting. This diagram is of the I/O port assignments.

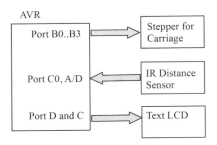

# COMBINATION LOCK OPENER USING A STEPPER

This machine tries all the combinations of a pad lock. It uses a stepper motor to move the dial. In the end, the machine was able to find the combination to the lock. One difficulty was mechanically connecting the stepper to the lock. One elegant way to solve this is to design and print a connector using a 3D printer. Another problem was that the motor and lock moved at different increments or steps. 10 steps of the motor equaled 3 clicks on the lock. After the 3 number combination is dialed on the lock, a DC motor tried to pull the lock open. For testing purposes, the code can be started with a combination close to the solution. A text LCD can be added to let the user know what numbers are being tried. In the picture, the DC motor is positioned above the lock and the stepper is directly below the DC motor.

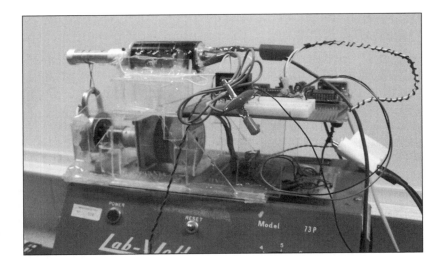

Use the following flowchart as a model of how to begin programming. The programming will be a little easier in C rather than Assembly.

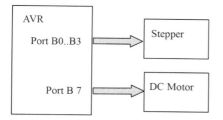

You will need to determine how many steps of the stepper are needed for 1 revolution of the lock. If the lock has 40 numbers per revolution, determine how many steps are required per 100 numbers. Call this value SPHN. The lock used in this example had 40 numbers per revolution. Before the program runs, manually dial the lock to the number just before the initial r16 value (25). In the flowchart, the program is to calculate the distance from r16 to r17. This is not straightforward. For example, what if r16 = 1 and r17 = 3 versus r16 = 39 and r17 = 1? Both are a distance of 2.

The flowchart is as follows:

```
Let r16=26, r17=17, r18 = 32     ;Numbers close to solution
Top:
   Call subroutine to check combination
   Increment r16.
   If r16>39 then r16=0 and increment r17
   If r17>39 then r17=0 and increment r18
   If r18>39 then end program, solution not found
   Goto top
Subroutine to check combination:
   Step   CW 2 revolutions - 1 number or (2*40-1)*SPHN/100 steps
   Step   CCW by (the distance from r16 to r17)*SPHN/100 steps
   Step   CW by (the distance from r17 to r18)*SPHN/100 steps
   Send bit to port to activate DC motor (pull on lock).
   End program if opened.
```

## SLEEP MODE WITH INTERRUPT

This example is more of a proof of concept as opposed to a project. The concepts laid out here demonstrate how to put the ATmega328 into a low power state which can be used in some project.

The idle sleep mode turns off the CPU and FLASH clocks but leaves on the I/O, ADC, and ASY clocks. The power down sleep mode turns off all clocks and oscillators. The idle mode uses more energy but can be brought up more quickly and from more wake up sources. Either mode can be woken up by putting a low level on INT1 or INT0 interrupt pins. There is also an ADC noise reduction sleep mode that can be used to improve the performance of the A/D converter. That sleep mode can be woken up by using the ADC interrupt complete. Select one of these first three lines of code to enter the appropriate sleep mode. Notice the last line of code uses in-line Assembly to enter the sleep mode.

```
SMCR = 0b0001;          // Idle sleep mode. Defaults to this.
SMCR = 0b0011;          // ADC noise reduction sleep mode.
SMCR = 0b0101;          // Power down sleep mode.
asm ( "sleep  \n" );    // Use this line to enter sleep.
```

So how much power does an ATmega328 use in these modes? Notice how little power the power down mode uses, 1/3600 that of the awake mode. The voltage regulator between the battery and AVR can use as much power as the AVR. Since the AVR can run on 3v to 5v, you can put three 1.5v batteries in series to get a supply voltage of 4.5v without a voltage regulator.

| State | Current (uA) | Power (mW) |
| --- | --- | --- |
| Awake, No Sleep | 1120 | 5.600 |
| Idle | 925 | 4.625 |
| ADC Noise Reduction | 885 | 4.425 |
| Power Down | 0.3 | 0.0015 |

In this code example, the INT0 pin is used to wake up the AVR from the power down sleep mode. INT0 is on pin 4 of the ATmega328. The '328 will wake up when INT0 is held low for a short period of time. Notice the interrupt service routine, ISR, and how an ISR is written in C.

```
// –––––––––––––––––––––––––––––––––––––––––––––
// atmega328_sleep.c
// Program shows how to enter the idle or power down sleep mode.
// It wakes up when an INT0 occurs.
#define LED 0b01

#include <avr/io.h>
#include <avr/interrupt.h>     // Needed for interrupts
#include <avr/sleep.h>         // Needed to use instruction: sleep_cpu();
```

```c
#include "delays_1mhz.h"

void led_pulse_long ();     // Turns LED on and off for 4 sec. each.
void led_pulse_short ();    // Turns LED on and off for 1/2 sec. each.

// --------------------------------------------------
int main(void)
{
   DDRB |= LED;             // Set port direction
   sei();                   // Enable interrupts
   EIMSK |= (1<<INT0);      // Allow interrupt INT0
                            // Pick one of the next 3 lines.
   SMCR = 0b0100;           // Power down sleep mode.
//SMCR = 0b0010;            // ADC noise reduction sleep mode.
//SMCR = 0b0000;            // Idle sleep mode. Defaults to this.
   SMCR |= (1<<SE);         // Allow sleep mode. Default mode is idle.

   while(1)                 // Loop again after waking up.
   { led_pulse_short();     // Indicates main program active.

     asm ( "sleep   \n" );  // Use this or next line to enter sleep.
//   sleep_cpu();           // Requires: #include <avr/interrupt.h>
                            // Sleeps stops main program

   }
} // -------------------------------------------------
// -------------------------------------------------
// Interrupt service routine.
ISR(INT0_vect)
{ led_pulse_long();         // Indicates ISR active.
  return;
} // -------------------------------------------------
// -------------------------------------------------
void led_pulse_long ()      // Turns LED on and off for 4 sec. each.
{ PORTB |= LED;
  delay_ms(4000);
  PORTB &= ~LED;
  delay_ms(4000);
} // -------------------------------------------------
// -------------------------------------------------
void led_pulse_short ()     // Turns LED on and off for 1/2 sec. each.
{ PORTB |= LED;
  delay_ms(500);
  PORTB &= ~LED;
  delay_ms(500);
} // -------------------------------------------------
```

# INDEX